The Man Who Planted Trees

The Man Who Planted Trees

LOST GROVES,
CHAMPION TREES,
AND AN URGENT PLAN
TO SAVE THE PLANET

Jim Robbins

Spiegel & Grau
NEW YORK
2012

Published in the United States by Spiegel & Grau,
an imprint of The Random House Publishing Group,
a division of Random House, Inc., New York.

SPIEGEL & GRAU and design is a registered
trademark of Random House, Inc.

LIBRARY OF CONGRESS CATALOGING-IN-PUBLICATION DATA
Robbins, Jim.
The man who planted trees : lost groves, champion trees,
and an urgent plan to save the planet / Jim Robbins.
  p. cm.
ISBN 978-1-4000-6906-4
eBook ISBN 978-1-58836-999-4
1. Forest germplasm resources conservation.
2. Forest conservation. 3. Tree planting. I. Title.
SD399.7.R63 2012
333.75'16—dc23   2011040738

This book was printed in the United States of America
on Rolland Enviro™ 100 Book, which is manufactured
using FSC-certified 100% postconsumer fiber
and meets permanent paper standards.

www.spiegelandgrau.com

987654321

FIRST EDITION

Book design by Debbie Glasserman

Without the following people this book would not have been possible, and to them it is dedicated: To David, who showed up with his wild stories, changed everything, and let me tell his story. To my editor, Cindy Spiegel, and bless her Job-like patience. For Chere, whom I love. For Matthew Robbins and Annika Robbins, who will inherit the future we leave. For Leslie Lee, Diana Beresford-Kroeger, Bill Libby, Bill Werner, Meryl Marsh, Kerry Milarch, Jared Milarch, Jake Milarch, Michael Taylor, Marybeth Eckhout, Chris Kroeger, Andrew St. Ledger, and all those who are trying to make the world a better place. Thanks to Stuart Bernstein, my agent, who believed in this book, and Donna Dell Dowden, for the beautiful drawings. Hannah Berglund, thanks for your help and the crepes. And thank you, especially, to the trees, who do so much and ask for so little.

A fool sees not the same tree that a wise man sees.

WILLIAM BLAKE

Trees are the earth's endless effort to speak to the listening heaven.

RABINDRANATH TAGORE, *FIREFLIES*, 1928

Preface

In 1993 my wife, Chere, and I bought fifteen acres of land on the outskirts of Helena, Montana. The property was tangled with a dense ponderosa pine forest so thick it's called dog hair, and some of the stubborn old trees had lived well there for centuries, in rocky terrain, marginal soil, and cold temperatures. We installed a Finnish woodstove called a *tulikivi*, a mammoth dark gray soapstone box about six feet tall, in the middle of the living room of our new house. Tulikivis are highly efficient because the soft, dense stone mass around the firebox soaks up heat from a roaring fire and holds the warmth for 24 hours. Heat from the stone is radiant, softer and more pleasant than the heat from a burning fire. It's also clean—woodstove pollution comes from damping down a fire so that it burns slowly, which gives off a dense cloud of smoke. This stove burns hot.

Our forest and stove were the perfect marriage—we planned to thin the trees gradually and feed the fire with what we cut. We had more than enough wood, I figured, to last not only my lifetime, but my children's and grandchildren's. We were fireproofing our home as well, reducing the fuel in the woods around us in case a wildfire should blow up. Thinning would help open the dense forest for the wildlife that occasionally appeared. A moose ran through the yard one day, a black bear turned over the barbecue, a bobcat sauntered by. The eerie ululating howls of coyotes echoed at night and sent chills down my dog's neck, and she wailed from her bed in the front hall.

One day in 2003, as I was hiking through the woods to town, I saw my first "fader." A fader is a tree that has been attacked and killed by insects and is slowly fading from green to a reddish brown. In this case the perpetrator was the mountain pine beetle, a small dark insect that burrows beneath the bark and lays its eggs. Larvae hatch and eat the *phloem,* a thin, moist membrane under the bark that is a life support system for the tree. As the grubs gobble their way around the trunk, they cut the crown of the tree off from its source of water and from nutrients in the soil; the life in the tree ebbs, and it fades slowly from life to death. Alarmed at the outbreak, I quickly chainsawed down the red tree and stripped off the bark to expose the insects to the cold. But I noticed several other large trees that, while still green, had been infected. Having finished this one off, the bugs had flown to other nearby trees to begin again.

The appearance and efflorescence of these bugs paralleled a series of warmer winters the West has experienced in the last few decades. In the 1970s and 1980s, winter usually meant two or three weeks or more of temperatures 20, 30, or 40 degrees Fahrenheit below zero. Since the 1980s the temperature has dipped into 30 below territory only a handful of times and never for a long period, and it has never fallen anywhere near the record of 70 below set in 1954. The warmer wintertime minimum temperatures make a huge difference to the bugs. The larvae are well adapted and manufacture their own supply of glycol—the same chemical found in automobile antifreeze—as they head into winter. They easily survive temperatures of 20 and 30 below.

Bark beetles such as the mountain pine beetle are highly evolved and ruthlessly efficient—the beetle's Latin genus name, *Dendroctonus,* means tree killer. They carry a blue fungus from tree to tree in pouches under their legs, a super-rich food source that they plant in the host tree. As it grows, they feed it to their young. After the larvae feast, they turn into adults, each the size of a letter on this page, and hatch from the host tree by the thousands. Swarming to a nearby uninfected tree, they tunnel through the bark, and once ensconced, they send out a chemical mist called a *pheromone,* a scented request for reinforcements. They need to overwhelm the tree quickly in order to kill it. When their numbers are large enough, they send out a follow-up pheromone message that says the tree is full. Each infected tree produces enough new insects to infect five to eight other trees.

The telltale signs of an infected tree are ivory-colored, nickel-sized plugs that look like candle wax plastered on the trunks, which mean the tree is pumping out resin to try to drown a drilling bug. Sometimes a tree wins by entombing a beetle; far more often these days, the tree loses to the mob assault. The stress caused by warmer temperatures and drought makes it harder for the trees to muster enough resin to resist attacks.

A year after I saw the first fader, a dozen trees in our backyard forest were dead. Then things went exponential. A dozen dead trees turned to thirty, which turned to 150, dying far faster than I could cut them down and turn them into firewood. Finally, five years after the first infected tree, virtually all of our trees were dead. We threw in the towel and hired loggers to come in and cut our trees down.

Enormous machinery rumbled into our woods on tank treads. A piece of equipment called an "excadozer" grabbed a tree, cut it off at the base, and stripped the branches in less than a minute. I watched trees falling around the house, heard the angry hornet whine of chain saws, and looked below to see our forest piled up like Lincoln logs ready to be turned into paste. The tang of pine scent hung thickly in the air.

It's eerie when the tree reaper comes to claim your forest and renders the once living world around you stone dead. When the forest is gone and the sky opens up, it's disorienting, as if someone has removed a wall of your house. Broken branches, smashed limbs, and slivered stumps were all that was left; tank tread marks scarred the exposed forest floor. "It's like open-heart surgery," said my "personal logger," Levi. "You don't want to watch while it's going on, but in the end it's a good thing."

The loggers hauled the logs off on trucks and loaded them onto a freight train headed for Missoula, Montana, where they were sold to a paper plant, which chipped them and turned them into an oatmeal-like slurry to become cardboard boxes. It cost us about a thousand dollars an acre to cut and ship out our dead trees. They have little value as lumber because of the blue stain from the fungus the beetles inject. And there are so many dead trees in Montana these days that it's a buyer's market. The landscape looks postapocalyptic in many places.

The once forested hill behind our house is now a meadow. A perch on a grassy hilltop a short hike from my house reveals an entire mountain valley draped with trees, nearly all of which are dead and dying. All around Helena, in fact, the forests are mostly dead, a jumble of rust-red dead trees.

Trees and forests come and go all the time. Like fire, beetles are agents of the natural cycle of life and death; they break down trees so their nutrients can be absorbed into the soil and used by the next generation of trees. But the boundaries between what is natural and what is unnatural have broken down—and this outbreak and others in the West appear to be unnatural. Climate change, experts say, is partly a human artifact and has elevated the beetles into a much larger destructive role than they might otherwise play. Freed from their usual boundary of cold temperatures, they have broadened their range and claimed more territory, going higher in altitude and pushing farther north than ever. They used to hatch from their tree and fly to a new host during two weeks in July; now they fly all summer and part of the fall. They used to attack only trees over five inches in diameter; now they are eating trees no bigger around than my thumb. Although the growing season for forests is two weeks or more longer than it used to be, the amount of precipitation is the same, and so the trees must go longer with the same amount of water, which has stressed them and reduced their defenses.

As I researched and wrote stories for *The New York Times*

and other publications about this massive change in the woods, it dawned on me—I mean really dawned on me—that North America's cordillera, the mountains that extend from Alaska to northern New Mexico, and that include my patch of forest, were ground zero for the largest die-off of forests in recorded history. "A continental-scale phenomenon," said one shocked scientist. Suddenly climate change hit home.

Imagine a world without trees.

It was no longer an abstract question for me. The forests on my property, in the valley where I live, in my region, and in much of my state are dying. The speed and thoroughness of the die-off is stunning.

Take a minute to imagine if every tree around your home, your city, your state were to wither and die. What would the world be like? And it promises to become much worse if even the most conservative climate predictions come to pass. Few people are thinking about the future of forests and trees.

Credit where credit is due. This book began a decade ago with vital questions that I never heard asked until they were posed by a third-generation shade tree farmer who claimed he had died and gone to heaven and then returned. He was worried about the future of the global forest and told me that the trees that make up most of the world's forests are the genetic runts—the scraps left after the last few centuries of orgiastic clear-cutting, which selected out the biggest, straightest, and healthiest trees. There is a strong, urgent case to be made that this is true, that we are witnessing a kind of evolution in reverse, and that genetically our forests are a shadow of what they once were and may not be strong enough to survive on a rapidly warming planet.

For a long time I doubted that this farmer, a recovering alcoholic, former competitive arm wrestler, and street brawler who claimed a fantastic ride out of his body and back, really knew what was going on with the world's trees. And then forests in my backyard and state and region and continent started dying; not just a few trees, but trees by the millions. They are still dying. His ideas seem more valid every day.

Aldo Leopold wrote that "to keep every cog and wheel is the first precaution of intelligent tinkering." It's unimpeachable common sense that we have ignored. When it comes to our ancient forests, we have nearly wiped them out—more than 90 percent of America's old-growth forest is gone and is still being cut, and some 80 percent of the world's has vanished. Yet we have only begun—*only begun*—to understand the ecological role these forests play or what secrets might be locked away in their genes.

Forests hold the natural world together. They have cradled the existence of our species since we first appeared—trees and forests are the highest-functioning members of ecological society, irreplaceable players at the apex of the complex ecological web around us. They are ecosystem engineers that create the conditions for other forms of life to exist, on every level. As I worked on this book the Gulf oil spill occurred, and was widely covered as the black tarry substance fouled beaches and killed wildlife. The loss of our trees and forests, our ecological infrastructure, isn't nearly as dramatic; it's a quiet crisis. But the impacts are far greater.

. . .

THIS BOOK IS about several things. It is first and foremost about David Milarch, the aforementioned tree farmer and an interesting guy if there ever was one. It is about the longtime human love affair with trees and forests, a love affair that has ebbed and flowed but has been ongoing since the dawn of humans, since our ancestors first climbed into them to escape predators. It is a book about the science of trees and of forests, and about the unappreciated roles they play in sustaining life on the planet.

Trees are responsible for half the photosynthesis on land, taking in the energy from sunlight and transforming it to leaves, where that energy is usable by insects and mammals and birds. They are highly evolved water management specialists; a forest is a soft carpet on the landscape that allows a downpour to reach the ground gently rather than in a torrent, so that it can be absorbed rather than run off and can recharge groundwater. Trees feed oxygen and minerals into the ocean; create rain; render mercury, nitrates, and other toxic wastes in the soil harmless; gather and neutralize sulfur dioxide, ozone, carbon dioxide, and other harmful air pollutants in their tissue; create homes and building materials; offer shade; provide medicine; and produce a wide variety of nuts and fruits. They sustain all manner of wildlife, birds, and insects with an array of food and shelter. They are the planet's heat shield, slowing the evaporation of water and cooling the earth. They generate vast clouds of chemicals that are vital to myriad aspects of the earth's ecosystems and likely vital to our health and well-being. They are natural reservoirs—as much as a hundred gallons of water can be stored in the crown of a large tree. The water they release is part of a largely unrecognized water cycle.

Even viewed conservatively, trees are worth far more than they cost to plant and maintain. The U.S. Forest Service's Center for Urban Forest Research found a ten-degree difference between the cool of a shaded park in Tucson and the open Sonoran desert. A tree planted in the right place, the center estimates, reduces the demand for air conditioning and can save 100 kilowatt hours in annual electrical use, about 2 to 8 percent of total use. Strategically planted trees can also shelter homes from wind, and in cold weather they can reduce heating fuel costs by 10 to 12 percent. A million strategically planted trees, the center figures, can save $10 million in energy costs. And trees increase property values, as much as 1 percent for each mature tree. These savings are offset somewhat by the cost of planting and maintaining trees, but on balance, if we had to pay for the services that trees provide, we couldn't afford them. Because trees offer their services in silence, and for free, we take them for granted.

This book also delves into the long history of the sacred aspects of trees, a worship that has been around for as long as people have. There is a view among many to this day that trees were a gift from the creator, placed here to provide for our basic food and shelter needs. Across cultures and across time, trees have been revered as sacred, as living antennae conducting divine energies. "The groves were God's first temples," wrote William Cullen Bryant in his 1824 poem "A Forest Hymn."

Science has not, as we like to think, conquered every realm. And one of the places it has a great deal of work to do is with the trees. What an irony it is that these living beings whose shade we sit in, whose fruit we eat, whose limbs we climb, whose roots we water, to whom most of us rarely give a second thought, are so poorly understood. We need to come, as soon as possible, to a profound understanding and appreciation for trees and forests and the vital roles that they play, for they are among our best allies in the uncertain future that is unfolding.

Contents

The Man Who Planted Trees

CHAPTER I

Champion Tree

THE ORIGINAL BOOK TITLED *The Man Who Planted Trees* is a slim volume, just four thousand words; in fact, it was first published as a story in *Vogue* magazine in 1954. That book was written as a fable by a Frenchman named Jean Giono and has tapped a deep well in the human imagination, and since its publication in book form, it has sold close to half a million copies. Speaking in the first person, its unnamed narrator describes hiking through the French Alps in 1910, enjoying the wilderness. As he passes through a desolate, parched mountain valley where crumbling buildings testify to a vanished settlement, he comes across a middle-aged shepherd taking his flock out to pasture. The shepherd has one hundred acorns with him, and he plants them as he cares for his sheep. It turns out that the shepherd has planted more than a hundred thousand trees on this barren, wind-ravaged landscape.

Six years later, after surviving the front lines of World War I, the narrator returns to the shepherd's hut. He is surprised to see small trees "spread out as far as the eye could reach. Creation seemed to have come about in a sort of chain reaction. . . . I saw water flowing in brooks that had been dry since the memory of man. . . . The wind, too, scattered seeds. As the water reappeared so too there reappeared willows, rushes, meadows, gardens, flowers and a certain purpose in being alive."

As the years go by, the trees grow taller and the forest in the valley grows thicker, and a dying ecosystem is transformed into a thriving one. When the narrator returns for a third time, toward the end of the story, more than ten thousand people are living in the flourishing valley.

> Everything was changed. Even the air. Instead of the harsh dry winds that used to attack me, a gentle breeze was blowing, laden with scents. A sound like water came from the mountains: it was the wind in the forest. Most amazing of all, I saw that a fountain had been built, that it flowed freely and—what touched me most—that someone had planted a linden beside it, a linden that must have been four years old, already in full leaf, the incontestable symbol of resurrection.

Some experts say *The Man Who Planted Trees* is wishful thinking, that reforestation cannot effect the kind of transformation imagined in the book, bringing a barren landscape back to life and bringing harmony to the people who live there. Planting trees, I myself thought for a long time, was a feel-good thing, a nice but feeble response to our litany of modern-day environmental problems. In the last few years, though, as I have read many dozens of articles and books and interviewed scientists here and abroad, my thinking on the issue has changed. Planting trees may be the single most important ecotechnology that we have to put the broken pieces of our planet back together.

Take the growing number of emerging infectious diseases. Their connection to the natural world is one of the most revelatory things I discovered about how little we understand the role of forests. I learned that there is a surprising single cause that connects a range of viral diseases including hantavirus, HIV, Ebola, SARS, swine flu, and West Nile virus with bacterial diseases including malaria and Lyme disease. Rather than just being a health issue, these deadly diseases are, at root, an ecological problem.

To put it in a nutshell, the teams of scientists researching the origins of disease say that pathogens don't just mysteriously appear and find their way into human populations; they are the direct result of the damage people have done, and continue to do, to the natural world, and they are preventable. "Any emerging disease in the last thirty or forty years has come about as a result of encroachment into forest," says Dr. Peter Daszak, director of EcoHealth Alliance, a New York–based international NGO that is pioneering the field of conservation medicine. "Three hundred and thirty new diseases have emerged since 1940, and it's a big problem." Most of these diseases are zoonotic, which means they originate in wildlife, whether in bats or deer or ticks, which then infect people who live near the forest. It's believed, for example, that the human immunodeficiency (HIV) virus crossed the species barrier from monkeys to humans when a bushmeat hunter killed a chimpanzee, caught the virus from the animal, and brought the disease out of the jungle and into the world of humans. Fragmenting forests by building subdivisons in the oak forests of Long Island or logging in the mahoghany forests of Brazil degrades the ecosystems and exacerbates disease transmission to humans.

SO THIS BOOK is not just about planting trees. It is about the state and the likely fate of the world's forests as the planet journeys into a possibly disastrous century of soaring temperatures. Precisely what such rapid warming is doing, and will do, to the forests is unknown, but more virulent pests and diseases, drought, climate extremes, high winds, and an increase in solar radiation will likely take a steep toll on the forests.

We are beyond known limits, and traveling farther beyond them every day.

What will happen to the trees and forests? There is no formal predictive model because trees and forests have been poorly studied; there are no long-term data, and the world's forests are extremely varied and complicated. Despite the lack of data, it doesn't take an ecologist to imagine what could happen. Apparently, though, it takes a journey into another realm to come up with an idea about what might be done to save our oldest trees in the event the changes become catastrophic.

My journey into the world of trees started in 2001, when I

read an article about an organization called the Champion Tree Project. At the time, the group's goal was to clone the champion of each of the 826 species of trees in the United States, make hundreds or thousands of copies, and plant the offspring in "living archival libraries" around the country to preserve the trees' DNA. A "champion" is a tree that has the highest combined score of three measurements: height, crown size, and diameter at breast height. The project's cofounder, David Milarch, a shade tree nurseryman from Copemish, Michigan, a village near Traverse City, said he eventually hoped to both sell and give away the baby trees cloned from the giants. "Clones," in this case, are human-assisted copies of trees made by taking cuttings of a tree and growing them—an old and widely used horticultural technique for growing plants. Unlike a seedling, which may have only 50 percent of the genetics of its parent, a clone of a tree is a 100 percent genetic duplicate of its parent.

I have always been drawn to big old trees, and the idea of making new trees with the genes of champions was compelling. I proposed a story to *The New York Times* science section about the idea, got the assignment, and drove to Big Timber, Montana, not far from my home, to visit Martin Flanagan, a lanky working cowboy and tree lover who helped gather materials for Milarch's Champion Tree Project in the West. On a bluebird day in May, Flanagan drove me down along the Yellowstone River, bank-full and the color of chocolate milk, as the spring sun melted snow in the mountains. He showed me several large trees, including a towering narrow-leaf cottonwood. "This is the one I plan to nominate for state champ," he said excitedly, spanking the tree with his hand. "It's a beauty, isn't it?"

There wasn't much to the Champion Tree organization, I found out. It was mostly a good idea with a tiny budget, with Milarch and occasionally one of his teenage sons working out of his home in Michigan; Flanagan working part time in Montana, driving around in a beat-up pickup truck gathering cuttings; and Terry Mock, from Palm Beach, Florida, who was the director.

Over the next week I interviewed Milarch several times by phone, and he talked to me about the need to clone champion trees. "The genetics of the biggest trees is disappearing. Someone's got to clone them and keep a record. No one knows what they mean. Let's protect them so they can be studied in case they are important. A tree that lives a thousand years might know something about survival." I also interviewed several scientists who agreed that researchers don't know the role that genetics plays in the longevity and survivability of trees; it simply hasn't been assessed. Environmental conditions, including soil and moisture, are obviously critical as well. Two identical clones planted twenty feet apart might grow far differently. Almost all of them said, however, that in the absence of study, it's Botany 101 that genetics is a critical part of what's essential to a long-lived tree. If you want to plant a tree and walk away and have it live, it makes sense to plant a tree that is the genetically fittest you can find. The big old-timers have proven their genetic mettle; they are survivors. Or as General George Cates, former chairman of the National Tree Trust, put it to me, "You can bet Wilt Chamberlain's parents weren't five foot one and five foot two."

Dr. Frank Gouin, a plant physiologist and the retired chairman of the horticulture department at the University of Maryland, is a friend of the project and spoke to me in support of the notion of cloning. He had cloned a big tree himself, the legendary 460-year-old Wye Oak on Maryland's eastern shore. "These trees are like people who have smoked all their lives and drank all their lives and are still kicking," Gouin said. "Let's study them." And the way to perpetuate and study them, he said, is just the way Champion Tree proposes.

My story about Champion Tree ran on the front of the *Science Times* section on July 10, 2001, with several color photographs of various champions, and over the next few days other media picked up the story. After a flurry of interviews, including eleven minutes on the *Today* show, Milarch, flabbergasted at the reach of the *Times,* called me. "It put us on the map, big time," he said. "I can't thank you enough." He said he wanted to come to Montana to meet me and give me a gift of a champion green ash tree as a thank-you. Though I loved the idea of a champion of my own, professional ethics prevented me from accepting the gift. "Let's plant one on the Montana capitol grounds instead," he suggested.

Fine, I said, a gift to the state.

With the attacks of 9/11, the tree planting wouldn't come until the following year. On a warm, sunny June day, David Milarch came to my office in downtown Helena and introduced himself with a big hand. He is a jovial bear of a man, six foot three with broad shoulders and big arms. He looks like a lumberjack and was dressed like a farmer, in a short-sleeved snap-button shirt, jeans, and a plastic foam farm cap that said OLYMPICS 2002, and he carried a hard-shelled briefcase. There is a bit of Viking in him, not only in his outgoing personality and swagger but in his ruddy complexion, though the hair that is left is white. A small strip of wispy white beard didn't cover his ample chin. A belly spilled over his belt.

Milarch has the charm gene, and I liked him right away. A born storyteller, he laughs loudly and frequently, and he has a flair for the dramatic and a fondness for announcing things rather than just saying them. He is an expert in the use of compliments, but pours it on a little too thick sometimes. As we talked he flipped open his briefcase and pulled out a crumpled pack of Marlboro Lights, put one in his mouth, and, in a practiced move, lit a cigarette with one hand by leaving the match attached to the book, folding it over, and lighting it with his thumb.

Over lunch, I expected a chat about the science of big tree genetics. I was wrong. As we sat down at a local restaurant, Milarch began a story. In 1991, he told me, he died and went to heaven. Literally. A serious drinker, he had quit cold turkey. The sudden withdrawal of alcohol caused kidney and liver failure, and a friend had to carry him to the emergency room, where a doctor managed to stabilize him. The next night, his wife and his mother beside him, he felt himself rise. Not his body, he said, but his awareness—he could look down from the top of the room and see himself lying there. It was a full-blown near-death experience, a phenomenon also known as disambiguation, something, at the time, I'd never heard of. His consciousness, he said, left the room and soon passed through brilliant white light—"It was like a goddamn blowtorch!" he told me. On the "other side" he was told it wasn't his time, that he still had work to do on earth, and he needed to go back. When his awareness returned to his body, he sat up in bed, shocking his wife and mother, who thought he was dead.

The experience changed him—afterward he felt more alive and more present—and he understood, for the first time, he said, the importance of unconditional love. He appreciated his children and family more, and had a deeper connection to music and art. He felt more intuitive and more spiritual, even more electric, so charged that he couldn't wear a wristwatch or use a computer—they were affected by his body's electrical properties, which had been enhanced somehow. He wasn't perfect; there was still some of the old David there. But it existed along with this new part of him.

Months later, still adjusting to this new life, he was visited in the early morning hours by light beings, who roused him. The big trees were dying, they told him, it was going to get much worse, and they had an assignment for him.

In the morning he told his kids that the family had a mission—to begin a project to clone the champion of every tree species in the country and plant them far and wide. They were a farm family in the middle of what many call nowhere, a world away from environmental groups and fund-raising and politics and science. But the Milarchs were hopeful, naïvely so, and unaware of the obstacles that confronted them.

Lunch came and I was quietly incredulous. Was I really hearing this? I thought he was joking or spinning a yarn, but he said it all with a straight face. It was, to say the least, the most unusual origin of a science story I'd ever heard. I'd had no inkling of any of it during phone interviews. It didn't diminish the science, as far as I was concerned, because all the scientists I'd interviewed for the story said cloning trees to save genetics is a scientifically sound idea. Where people sourced their inspiration didn't matter if the science passed the test. Still, it was curious. And this chain-smoking tree farmer who liberally deployed the F-bomb didn't fit the mold of the typical New Ager.

During his visit over the next couple of days, Milarch laid out his take on what humans have done to the world's forests, based on his peculiar blend of science and intuition, and how the Champion Tree Project wanted to change that. "People should be awestruck, outraged, overwhelmed," he said. "A tree that is five, six, eight, or fifteen feet across, the champions we are cloning, is what the size of all the trees in our forests once was, that all of America was covered with, not just one lone, last soldier standing. When we look at the trees around us, we're looking at the runts, the leftovers. The whole country should be forested coast to coast with these giants, not with the puny, scraggly, miserable mess we call our forests. We don't realize what we've lost."

"The champions are in harm's way," he told me. "They do their best in communities that are hundreds or thousands of acres. They're struggling in little pockets to hang on. We either get on it and get it done, or in twenty years they'll be gone. If man doesn't take them out, Mother Nature will. We're in the fifty-ninth minute of the last hour."

"Why do these light beings care about trees?" I asked.

"They are concerned about the survival of the planet. Call them light beings, plant *devas*, earth spirits, or angels, they are real, and there are some in charge of the trees. Americans are about the only ones who don't believe in such things, but they are out there and a lot of people can hear them, including me. We treat the earth like it's dead, which allows us to do what we want, but it's not dead." Genetics is critical to the survival of the forests, they told him, and one day science will be able to prove it.

He has taken his lumps from critics, including from some scientists who say he doesn't know about science, and that it's not known if genetics is what make these trees survivors. They say it could be just plain luck that these trees have survived. It's not a bad idea to clone the trees, they say, just not necessary. Others have said that it is impossible to clone a two-thousand-year-old tree—that it's akin to asking a ninety-year-old woman to give birth.

"They can criticize all they want. But these are the supertrees, and they have stood the test of time," said Milarch. "Until we started cloning the nation's largest and oldest trees, they were allowed to tip over, and their genes to disappear. Is that good science? If you saw the last dinosaur egg, would you pick it and save it for study or let it disappear?"

What we have lost by mowing down the forests around the world, he insists, is far more than big trees. We've squandered the genetic fitness of future forests. With humans high grading, cutting down the best trees time and again, the great irreplaceable trove of DNA that had been shaped and strengthened over millennia by surviving drought, disease, pestilence, heat, and cold—the genetic memory—was also gone. The DNA that may be best suited for the tree's journey into an uncertain future on a warming planet has all but vanished, just when we need it most. By making copies of the cream of the big tree crop and planting

thousands or tens of thousands of the 100 percent genetic copies of these "last soldiers standing" all over the country or the
world, Milarch wants to ensure that they will live on. They may
well not turn out to be better, but if they do turn out to have superior traits, then their genetic traits have been saved—"money in
the bank," Milarch says. When it comes to trees, proven survi-
vor status is all we have to go on. Trees cradle human existence
in ways known and many more ways unknown, he said. "And if
we lose the trees, we're in a world of hurt."

There are other reasons to plant trees, of course. It's a way
to address climate change by soaking up carbon dioxide. Los
Angeles, for example, has a campaign to plant a million trees;
if it reaches its goal, because each tree will store two hundred
pounds of carbon annually, that will be the equivalent of taking seven thousand cars off the road every year. Trees also filter
air pollution, water pollution, and toxic waste in the soil. Flooding, though often thought of as a natural catastrophe, is more
often than not man-made, caused when water is unleashed
from the natural storage and regulation of forests and marshes,
exacerbated by deforestation. A one-hundred-year flood event
becomes a one-in-five-year event when deforested land cover is
a quarter of the total. The floods caused by deforestation also
last longer—4 to 8 percent longer for each 10 percent of the forest that is lost. And the floods are more severe than when forests
are intact. Ninety percent of the natural disasters in the United
States involve flooding, and floods have become far more frequent, largely because of deforestation. Recent unprecedented
floods in Pakistan may have been in large part the results of a
warmer planet—2010 was the warmest year on record, and a

warmer atmosphere holds more moisture. But Pakistan also has one of the highest rates of deforestation in the world. Between 1990 and 2005 it lost a quarter of its natural forest cover. An irony is at work here: as the climate grows warmer, the atmosphere will carry even more precipitation. Already, scientists say, rainstorms and snowfall have become more extreme. The increased precipitation around the globe in the coming years will occur when the landscape is least equipped to handle it.

"It seems hopeless," Milarch said. "The planet is warming and we hear this bad news all the time. We decided to do something. With no money and no staff. Some of the experts told us it couldn't be done. But we couldn't just sit here and watch the ship go down. And we are doing it. And we hope it's an example to other families." Someday, Milarch went on, he envisioned a movement, people planting clones of a global collection of thousands of the world's great trees, from the cedars of Lebanon to the ancient oaks of Scotland, Ireland, and England—not only to restore forests, but to create functioning old-growth forests in cities and suburbs around the world. Living among them, we would be able to take advantage of their ecosystem services, their ability to cool hot asphalt, to protect us from ultraviolet radiation, to cleanse the soil and air of pollutants, to grow nuts and fruits, to calm us and beautify our surroundings.

It was such big talk for a farmer who had no funding, even if he had died and gone to heaven. He wanted me to write a book about his inspiration, but I couldn't see it. There's a saying in Montana: Big hat, no cattle. How would he ever bring his fascinating idea to fruition with no funding? Besides, I'm a science reporter, and I had trouble coming to terms with a wild story about out-of-body travel to heavenly realms and back. He was asking me to believe in a magical universe, and it was a big step to take over lunch.

While I didn't see a book in his tale, Milarch certainly had my attention and started me thinking. There was something in what he said, and the way he said it, that I couldn't dismiss. This was in 2001, before my own forest had started to die. Could he be right about forest die-off? Are the world's trees in big trouble? Is old-growth genetics important? I had written a great deal about forests in my thirty years as a journalist, but never with these ideas in mind.

Not long after David's visit, I joined classes from a local high school in a small ceremony on the front lawn of Montana's capitol. As we planted a champion green ash tree that the Champion Tree Project had sent by truck, the marching band played the theme from *Mission: Impossible*.

CHAPTER 2

White Oak

IN 1996, DAVID MILARCH called Frank Gouin to see if the Champion Tree Project could obtain a clone of the venerable Wye Oak, the country's oldest and largest white oak tree, named for the nearby town of Wye Mills, on Maryland's eastern shore. At the time, it was just under 32 feet in circumference, nearly 7 feet in diameter, and 96 feet tall, and the crown of the tree covered a third of an acre.

Dating to the sixteenth century, the Wye Oak witnessed the changes of history. The Choptank Trail alongside it grew from an Indian footpath through the woods to a wagon road to a highway. Because of its size, the tree was a local landmark as far back as the mid-nineteenth century. The Wye Oak got official recognition in 1909, when Maryland's state forester, Fred Besley—who also created the Champion Tree measurement formula and the first list of champions in 1925—measured and photographed it and later named it one of the first champions on his list.

Visitors started coming by to see it. In 1939, the Maryland General Assembly, recognizing the tree's historic and champion status, purchased one acre of ground around it to create a park. As the old oak started to decline, foresters for the state of Maryland made a valiant effort to keep the tree alive. Its base was hollow, and they dropped cyanide pellets into the cavity to kill termites and carpenter ants; they supported the branches with long strands of thick steel cable to keep them from breaking. "Man, it was a big tree," said Gouin. "Four guys could sit around a table and play cards in the cavity of that tree."

As a horticulturalist and tree lover, Gouin took note as state foresters tried repeatedly to clone the Wye Oak. "Nurserymen have always had a tough time grafting oak trees because they are so slow-growing," said Gouin, referring to the way that the cells on the tip and root stock combine slowly. "The state had spent some twenty-seven years trying to clone the Wye with grafting and budding and tissue culture. They were not able to, though they had tried everything."

But Gouin, with a thick shock of white hair and an Amish-style beard, is a master gardener who grows Christmas trees, persimmons, and peaches, and when Milarch called he was moved by Milarch's notion of cloning each of the champions. He also felt there was something special about the Wye Oak, and that those traits were likely stored in the genes. "A lot of white oaks get blight, but I have never seen blight on this tree," he said. "No oak wilt and no gypsy moths, either."

Trees are most often cloned by taking a six-inch or so cutting of the newest growth. A small opening is scraped into the cambium layer at the base of the cutting. It's then immersed in a mix of soil and hormones, and if the procedure is successful, new roots emerge from the small scrape. But Gouin had a different idea. Over a period of three years, he took three hundred cuttings of scion wood from the tree each spring. Scion wood is part of a branch that has several buds on it, and Gouin grafted those cuttings onto the roots of seedlings from the Wye Oak, which had been grown from acorns and so had inherited at least 50 percent of their genetics from the parent Wye and 50 percent from another tree, through pollination. Using such closely related root stock, Gouin figured, would be comparable to a bone marrow transplant among family members and would reduce the possibility of rejection. He used a tight rubber band to join the roots and the tips together. Getting the graft to retain moisture long enough for the slow-growing oak tips to successfully connect with the root was critical, so he wrapped them in tape and then buried them in soil.

The first year, just two cuttings of the three hundred taken were successfully grafted. The second year, he took fresh cuttings and nine clones survived. The third year, in 2002, Gouin visited the clones more often, making the 140-mile round trip to the nursery once a week to ensure that the soil was kept consistently wet and cool, and thirty clones of the Wye Oak grew. "I finally figured out what I was doing," said Gouin. Subsequent years, he felt, would produce even more clones. On June 6 of that year, however, a violent thunderstorm roared across the eastern shore and toppled the Wye Oak. Gouin drove straight to the fallen tree to take more fresh cuttings before the branches perished, but it was the wrong time of year. "We couldn't get any more good budwood," he said, "and none of them took."

After the oak fell, thousands of people from around the region went to pay their last respects to the fallen champion, mournfully taking pictures and gathering pieces of the tree. "We all kind of hoped the Wye Oak would never fall down," said Maureen Brooks, coordinator of the Maryland Big Tree Program.

Two of the Wye Oak clones were given to Mount Vernon for an Arbor Day planting, and they now grow there on George Washington's estate. Another clone is planted inside the stump of the old Wye Oak and is producing acorns. The other clones are in a seed orchard operated by the State of Maryland Department of Forestry and are sexually mature. "The clones are already producing acorns," said Gouin, "and you can buy a seedling from the state of Maryland."

The Wye Oak may be gone, but its genetic heirs live on.

CHAPTER 3

Forests on a Warmer Planet

IN 1982, A controversial and cryptic visionary named John Allen, along with his followers, built a $200 million structure called Biosphere 2 in the saguaro-studded Sonoran desert north of Tucson. The glass-and-steel-tube dome and other buildings, the size of two and a half football fields, are a contradiction in terms: a man-made natural environment that includes a mini-savanna, desert, ocean, coral reef, and tropical rain forest. The vivarium, as the complex of buildings is known, was meant to be a self-sustaining mini–Planet Earth that would function like the real Earth, with no external input.

Biosphere 2 was a pilot project for the day Allen and his followers hoped they could fly the pieces of the vivarium to Mars, assemble it there, and allow the group to "terra-form" the planet to create a human habitat. One by one, they thought, biospheres flown to Mars could be used to create a reservoir of trees, flowers, and other forms of plant life until one day the planet would be covered in vegetation and would sustain life as Earth does. The project collapsed, however, and the University of Arizona took over Biosphere 2.

While it didn't colonize the red planet, Biosphere 2 has proven nifty for studying planet Earth because it allows researchers to control variables and play out different scenarios in a way that can't be done in the real world. In 2009, two researchers, Henry Adams of the University of Arizona's ecology and evolutionary biology department and his professor David D. Breshears of the University of Arizona's School of Natural Resources, moved twenty mature piñon pine trees, five to six feet tall, into the dome and split them into two groups of ten. One group was placed in a chamber where conditions were equal to what they are today, and the other group was placed in conditions some seven degrees Fahrenheit warmer than it is now—roughly equal to the high end of the temperature rise scientists predict for the next century. Once the trees established themselves, researchers deprived both populations of water.

The human-induced drought killed the trees in the warmer chamber 28 percent faster than the trees in the chamber with normal temperatures. The message from the mini-Earth, researchers say, is that forest die-offs could increase by a factor of five if the climate warms seven degrees as predicted over the next century. "Droughts can kill trees faster when the temperatures are warmer," Breshears said. "Instead of one die-off in one hundred years, that number could increase by a factor of five. That's the take-home message. When I saw that in the data I went 'wow' and 'yikes.' This looks like a big deal, and it's based on temperature alone."

Based on temperature alone. That means the study looked only at the effect of warmer temperatures on trees, and not at the effect of warmer temperatures on disease and insects as well, which in concert with water stress kill many more trees than drought alone. And it makes the very conservative assumption that the increase in forest die-offs as temperatures increase will be linear rather than curvilinear. "Curvilinear means it gets a little warmer and it will kill a few more trees," said Breshears. "Then a little warmer, only this time it kills a lot more trees and perhaps a lot more species of tree."

The most recent forecast for the mean global temperature rise by the end of the century is 6.3 degrees Fahrenheit. That's a huge change in a very short period of time. Since the last ice age, twelve thousand years ago, the average global temperature has stayed fairly steady. If the global average increases as forecast, the changes will be wrenching, and the climate won't just get warmer. James Hansen, who heads the NASA Goddard Institute for Space Studies, writes in his book *Storms of My Grandchildren* that under present predictions, the earth is on track to see 20 percent of its species become extinct or be on their way to extinction by the end of the century, and to see droughts, heat waves, and forest fires of unprecedented ferocity, rapidly rising sea levels, and more storms with hurricane-force winds. E. O. Wilson, the distinguished professor of biology at Harvard, believes that by 2100, *half* of all plant and animal species could be extinct. Some leading scientists think we are in the beginning stages of a new mass extinction, one of six in

Earth's history. This one is different from the others, they say, because it is brought on by humans through the fragmentation of the natural landscape, the introduction of exotic species and new pathogens, the alteration of the climate, and unsustainable exploitation of plant and animal species.

Here's one of the ways warmer temperatures impact trees. A critical part of a tree is a cell found in its leaves and needles, called a *stoma*. Each of these pores is surrounded by two guard cells that act as a kind of valve, opening and closing the stoma according to changes in light, humidity, and the concentration of carbon dioxide. The tree takes water up from the ground through its roots and transpires it—that is, it releases water vapor into the atmosphere—through the stomata. As the water makes its way up to the canopy, it supplies not only moisture but also minerals to the entire tree.

But along with releasing water, the stomata also take in carbon dioxide for photosynthesis. So when they open, it is at the cost of releasing the tree's water stores. The death of the trees in Biosphere 2 was due to something called *carbon starvation*—at least, that's the hypothesis, because precisely how an individual tree dies is not clear. Researchers do know that when a drought hits, a hormone released by the tree tells the stomata to close in order to preserve the dwindling water supply in the air spaces of the leaf. With its stomata closed, though, the tree can no longer take in carbon dioxide, and so photosynthesis stops, and it must use its carbon stores, the food stashed in its pantry. When those stores are gone, the tree perishes. What Breshears and Adams hypothesize is that warmer temperatures cause trees to use up their carbon stores faster. "The tree says, 'I am going to close my stomata and ride this out,' " Breshears explained. " 'I am not going to do anything until I get more water.' But you are paying a cost every day, and at some point you collapse because you run out of juice."

THERE'S A STORY about a drunk searching for his car keys under a streetlight. Another man walks by and asks him what he is doing. The first man says he is looking for his car keys, and the second asks where he dropped them. "Over there," the first man says, gesturing toward the dark. "But there's more light here."

It's an apt metaphor for science. Many important subjects are difficult to study, or generate little profit, or aren't popular, and so they are largely ignored. So we assume implicitly they are not important. Much about trees and forests, it turns out, falls outside the ring of light.

Virtually every scientist I interviewed for this book said the same thing: we are hamstrung in assessing the health and risks to forests by an overwhelming lack of data and knowledge. So if we are waiting for a team of distinguished scientists to issue a report warning that the world's forests are in crisis and recommending a ten-point plan to fix them, forget it. The future of the world's forests is a black box and no one can reliably predict what is going to happen on a stable planet, let alone on one as unstable as ours is forecast to become. And the impact to the world's forests may already be well under way.

The giant forest die-offs in the Rocky Mountain West that my small forest tract is part of exemplify the secondary impact of warmer temperatures that Breshears referenced—in my case, bark beetles—and those secondary aspects may be far more deadly and unpredictable than heat and drought alone. In the last half century, the two-degree rise that has occurred in the West has already begun to turn ecosystems inside out, and the anomalous behavior of insects is one of those changes.

Dr. Diana L. Six is an entomologist at the University of Montana, and she knows well the biological amplification warmer temperatures bring. As we toured a beetle-killed forest, buzzing along dirt logging roads in the Blackfoot Valley, a half hour's drive out of Missoula, she explained how things have changed in her neck of the woods. "I used to plan my field season for the same two or three weeks in July," she said, which is when her quarry, tiny black mountain pine beetles, hatched from the tree they had just killed and swarmed to a new one. "Now everything is different. Instead of just two weeks, the beetles fly constantly from May until October, attacking trees, burrowing in, and laying their eggs for fully half the year." There are other anomalies. The beetles rarely used to attack immature trees, and now they do so all the time. And there's evidence that the beetles are switching to other species of tree, which has been unheard of. In some high places where the beetles had a two-year life cycle because of cold temperatures, it's changed to a one-year cycle, which means that their populations can increase more rapidly during warm droughts when their host trees are most stressed, resulting in more beetles doing a lot more damage.

In some ways, Dr. Six allowed, it's an exciting time to be an entomologist as the field's textbooks get rewritten. But it's also deeply unsettling. Large-scale die-off sometimes happens in a forest, she knows, but this magnitude is surreal. "A couple of degrees warmer could create multiple generations of beetles a year," she told me as she chopped a piece of bark off a dead lodgepole pine to show me the galleries of burrowing larvae, shallow squiggly channels in the wood. Trees here are the standing dead, varying shades of gray and brown, their bark drying and falling to the ground in jigsaw-puzzle pieces. "If that happens, I expect it would be a disaster for all of our pine populations." She looked around for a minute, more than a little aghast at how the ordered world of her science has been upended. "The whole ecosystem is changing," she said. "I've never seen anything like it."

As of 2010, across Colorado, Wyoming, Idaho, and Montana, about eight million acres of lodgepole and ponderosa pine had been killed. That's small change compared to what has happened recently in the Canadian province of British Columbia. The second largest known die-off occurred there in the 1980s and claimed a million and a half acres. The latest has claimed 43 million acres. By the time the outbreak ends, British Columbia may lose 80 percent of its old-growth lodgepole pine trees. Moreover, a freak wind event a few years ago blew the insects across the Continental Divide for the first time in history, and, freed from the bitterly cold temperatures that once kept them in check, they could, Canadian officials fear, munch their way across Canada's boreal forest, which forms a crescent across the northern half of the country, all the way to the Atlantic Ocean. This insect damage to Canada's forest comes on top of the devastation of the mature forests of British Columbia by logging. The only thing harder on trees than beetles, it seems, is people.

Dr. Craig D. Allen is a forest ecologist with Jemez Mountains

Field Station, part of the U.S. Geological Survey, the federal agency that studies the earth's natural systems. He's one of the scientists at the center of the study of climate-related forest changes and is a core member of the USGS Western Mountain Initiative, a coalition of six federal agency scientists and numerous university colleagues who are studying mountain ecosystems across the western United States and how these unique ecosystems are responding to a warming climate. He also works on forest die-offs in other parts of the world. Wiry and intense, with short graying hair and glasses, Allen has memorized a global list of troubling die-offs, and he hits the highlights as we walk through the Ancestral Pueblo ruins near his office in Bandelier National Monument. In 2005, for example, a one-year climate event in the Amazon, an El Niño drought accompanied by high winds, killed large numbers of trees. This event is well documented because so many researchers had study plots on the ground for other research projects. Moreover, during the drought, a violent storm swept through the region, and according to experts from NASA and Tulane University, that single storm killed a whopping half a billion trees in forty-eight hours. That year, the Amazon Basin went from being a carbon sink to being a source of carbon to the atmosphere. In 2010 another, even more severe drought struck.

But to see part of one of the worst cases of forest die-off ever documented, Allen needed look no farther than out his office window. In crystalline sunshine, as only the Southwest can do it, we walked across a mesa-top of dead, gray needleless piñons. This is the southern end of the sprawling patchwork of die-off that stretches from here, a couple of hundred miles north of the

Mexican border, all the way to Alaska—and includes the trees in my backyard forest. Secretary of the Interior Ken Salazar has called the die-off in the Rocky Mountains the West's version of Hurricane Katrina.

The die-off around the Southwest is attributed to a five-year drought that struck in 2000 and killed 90 percent of the trees in some areas, with widespread mortality across millions of acres. Populations of many species of bark beetles exploded (the Arizona pine engraver, pinyon ips beetle, and the Southwest pine beetle among them) and attacked many different tree species, presumably because the forests were stressed by drought, although this is one of the many things that researchers don't know for sure. And it's a conundrum that gets to the heart of attempts to assess forest die-offs.

It would seem a no-brainer to say that this die-off is caused by a warmer climate. In the last half century the average temperature in the Rocky Mountain West has gone up some two degrees, but the wintertime minimum temperatures—the really cold below-zero temperatures that kill overwintering beetles—have soared fifteen to twenty degrees, allowing more insects to survive. But there are so many other factors—fire suppression that has resulted in too many trees, for example, and a lack of research in key areas—that scientists can't say for certain that there's a direct link. "There are huge information gaps and uncertainties," Allen told me. "Huge. We don't even know for sure exactly how a tree dies. Not only can we not predict mortality, we can't even tell you for sure if forest mortality is increasing globally in recent years, as there hasn't been consistent long-term monitoring." If scientists could predict die-offs in the same way that climate forecasters can forecast rising sea levels, land managers could take preemptive action, such as mechanically thinning dense forest, or prescribing fires to reduce tree numbers. With fewer trees, moisture goes further, which increases the vigor of stressed forests, and it might allow them to fend off insects and disease. Even those actions, though, may just buy time as temperatures warm.

A dying forest is problem enough, but when large landscapes die they become contributors to a warming climate, a cycle called a *positive feedback loop*. One of the most important things forests do for life on the planet is to capture carbon dioxide from the atmosphere and store it as plant tissue; trees contain half of the terrestrial stores, more than any other single source on land. Otherwise the carbon dioxide would still be in the atmosphere, causing more warming. When forests die or are cut down, they release that stored carbon dioxide into the atmosphere; in fact those sources comprise 20 percent of annual carbon emissions. So much forest has died in British Columbia that, just as in the Amazon, the province went from being a net carbon sink—it sequestered more carbon than it emitted—to a carbon source in 2008. More global forest die-offs and their carbon releases compound warming, which can cause more forests to die, and around it goes.

Another example of a cycle that is changing weather patterns to create drier conditions involves heavy isotopes found in water. Mature trees take up as much as four hundred gallons of water per day, more than they need, and release much of it into the atmosphere, and the moist exhalation of the forests moves with the prevailing winds. Without this forest phenomenon, much of the interior of countries would be desert. It's difficult to research, but some new studies shed some light on how coastal forests create optimum growing conditions for forests a great distance away. In the Amazon, researchers found that not all water is created equal. One in every 500 molecules of water they examined had a heavier than normal hydrogen atom, and one in every 6,500 water molecules had a heavy version of the oxygen atom. While the heavier molecules are much slower to evaporate from surface water, the researchers found that these molecules were readily pumped into the atmosphere through transpiration, the release of moisture into the atmosphere through the leaves of trees. Forests, in other words, are key to keeping these heavy molecules moving through the water cycle. Research shows that in the last few decades, the number of heavy molecules has significantly declined, because so much of the forest has been cut down. "With many trees now gone and the forest degraded, the moisture that reaches the Andes has clearly lost the heavy isotopes that used to be recycled so effectively," scientists reported. That has markedly reduced the amount of rainfall.

Trees are also home to an unusual microorganism that may be a major contributor to rainfall, something scientists are just beginning to understand. Dr. David Sands, a researcher at Montana State University, and several colleagues have identified bacteria called *Pseudomonas syringae* that make their home on the leaves of most vegetation. The research team believes that as these bacteria are swept into the atmosphere and absorbed into clouds, a protein in them forms the nucleus for rain. When they fall to earth in rain and snow, they find their way back onto the leaves, where they multiply and start the cycle over again.

IN SPITE OF the lack of data and research, many scientists, Six and Allen among them, believe something unprecedented is happening to the world's forests. In Colorado, for example, not only have the mature lodgepole pines nearly been wiped out, but the quaking aspen, a deciduous tree famous for turning the mountain slopes a brilliant yellow in the fall, is also in danger. In 2005, researchers noticed that the aspens were dying in large numbers; thirty thousand acres perished that year. A hundred fifty thousand died the next year, and the number soared to more than half a million acres out of a total of about five million acres the following year. Not only are mature aspen stems dying, but the root mass underground, the source of all the trees aboveground, is dying. The rapid decline slowed in 2010, but experts say it was probably because of a couple of wet winters, and they fear the die-off could return.

Diana Six told me that she suspects that hotter, drier weather and a shift in rainfall patterns are responsible for unusual mortality rates in other places where she works. "Whole hillsides are suddenly dropping dead in South Africa," she told me. "It's happening so fast people are in shock. It's a tragedy." The dying species include the quiver tree—so named because it's fashioned by tribesmen into quivers to carry arrows—the camel thorn, and the giant euphorbia, a beautiful, weirdly shaped thirty-foot-tall succulent that looks like something out of a Dr. Seuss book.

"The feeling about forest die-off," said Allen, who is in touch with researchers around the globe, "is that there's a lot of it and it's flying under the radar." He can make that statement as a person who knows a lot about trees and forests, but as a scientist armed with nailed-down, bulletproof fact, he cannot. It's a painful situation to believe that there's a climate change crisis in the forests and not to be able to sound the alarm. "These recent die-offs we are seeing are before we put two to four degrees centigrade of warming into the system," said Allen, referring to the prediction for the rest of this century. "We don't know how much stress the forests of the world can take."

CHAPTER 4
Bristlecone

BRISTLECONE PINE FORESTS are found on mountaintops in Colorado, Utah, Nevada, California, and New Mexico. The oldest of the bristlecones, named for the bristlelike spine on their cones, are nearly five thousand years old, almost twice the age of redwoods and sequoias, yet a fraction of their size. The tallest specimen rises to just sixty feet, and most are much shorter. Not only are they the oldest trees in the world, they are among the world's oldest living organisms.*

The Great Basin bristlecone pines that grow in Great Basin

*Bristlecones are the oldest trees, but there are root masses below ground, or "clonal colonies," from which much younger trees grow on the surface. The known oldest of these is Pando, or the "Trembling Giant," an eighty-thousand-year-old aspen root mass in Utah and, at 6,600 tons, the world's heaviest organism. Although some might consider Pando the oldest living tree, when discussing trees I am referring to those lone trees that suffer the slings and arrows of the aboveground environment.

National Park in eastern Nevada are one of three species of the tree. The Wheeler Peak Grove there, in a mountain cirque, is one of my favorite groves of any species. The trees are gnomish, twisted into bizarre shapes like sculptures or bonsai trees, an adaptation to life in this harsh part of the world. The soil is rocky and alkaline, with few nutrients, and the winters are ten months long. The trees are exposed to high winds in this aerie, and their crowns often lean way over, while others hug the ground, a response called *wind training*. The bristlecone has survived in this unforgiving environment with an unusual strategy: as a main root dies, the section of trunk above it dies as well. But the tree keeps a small strip of living tissue alive. Until the climate started warming, the ability to adapt was its greatest strength, a strategy its enemies could not match.

Wheeler Peak is the grove where Prometheus fell, a bristlecone tree named by a group of local researchers and tree enthusiasts in the 1950s. It was "stooped as under a burden, with roots like claws grasping the ground, a magnificent monster standing alone," wrote a local author named Darwin Lambert, who helped name it, in a piece for *Audubon* magazine. "Four spans of my outstretched arms, six feet to the reach, were needed to encircle the misshapen trunk." They named it after the Greek demigod who was immortal and chained to a mountain crest by Zeus for aiding humans with the gifts of fire and art.

In 1964, a graduate student in geography from the University of North Carolina at Chapel Hill, Donald Rusk Currey, came to Wheeler Peak to study climate records in tree rings. Because bristlecones are so long-lived, they harbor an unparalleled record in their thousands of rings. There are conflicting accounts about what happened next, but apparently Currey drilled into Prometheus—whose name he did not know and whom he called WPN-114, for it was the 114th tree he sampled—to withdraw a core sample, thinking it was very old and held a long record of climate. Unfortunately his drill got stuck. Without it, he couldn't continue his research, so he cut down the bristlecone to retrieve the tool. Currey carried a slice of the bristlecone back to his motel and sat outside in the desert sun to count its rings. He finished at 4,844, but he believed the tree to be even older because the section wasn't from the bottom of the tree, which contained additional rings. Prometheus, he estimated, was at least 4,900 years old, perhaps 5,000, when he abruptly ended its life. Others believe it was 5,100 years old. It was the oldest known tree in the world.

Lambert was heartsick when he read Currey's paper in the journal *Ecology* and realized what the scientist had done. "The wounds open every time—to this day—when the memories of that ancient tree surface," he wrote later. Prometheus became a martyr to the cause of bristlecone protection. Lambert used the story of the destruction of the tree to campaign for national park status for the grove, and in 1986, Great Basin National Park was established.

Now, almost five decades later, and after five millennia dwelling at the roof of the world, all of the bristlecone pines may soon meet their match—though the threat is not the chain saw. The study of tree rings from the ancient trees shows that in the last fifty years, temperatures in the lair of the bristlecone are warmer than they have been in any other fifty-year period in the last 3,700 years. And that means the frigid temperatures that once protected the trees by killing off threatening insects are disappearing and the trees are becoming more defenseless.

Compounding the attack of the bark beetles is blister rust, a fungal disease that came to the West Coast from Asia, via Europe, on a shipment of white pine seedlings in 1900, spread across the West, and a hundred years later has reached the most remote and inaccessible mountaintops—probably because the climate has warmed. Because blister rust kills young trees quickly, and the pine beetle kills older trees, which produce seeds, the future of the bristlecone appears grim—in fact, over the coming decades, scientists say, all of the ancient bristlecones in the West will likely die out. So scientists are searching for trees that are resistant to the rust and planting seeds from those trees to grow new bristlecones, which they say offers the best chance for survival of the species. The case of the bristlecone illustrates the importance of preserving a range of genetics in a changing world. But while a new crop of the resistant pines will carry on the species, the oldest trees in the world will most likely not survive.

The loss of the bristlecone might represent more than the loss of a single species; it could be a loss to the entire ecosystem. The bristlecones and other mountaintop pines are what are known as *foundation species*—they create conditions that allow other species to gain a foothold, by providing habitat, food, and shelter. They build small pockets of soil in between their roots where other trees can grow, and they slow the melting of snow with their shade, which causes a slower release of spring runoff.

The bristlecones are also a window into longevity. They are members of an exclusive club of organisms on the planet that defy senescence—they don't appear to age, or to age very much. Even at several thousand years old they reproduce as if they were still teenagers. No one knows why or how they live so long. The question, sadly, is whether they will live long enough for us to find out.

CHAPTER 5

Great Unknowns

DAVID MILARCH WAS born in Detroit on October 5, 1949. His great-grandparents had emigrated from Germany and Ireland to Michigan, where they cleared forty acres near Copemish to grow food for a family of ten. After World War II there was more money to be made in the city, so the family decamped for suburban Detroit. His father, Edward, owned E. L. Milarch Nursery, where he grew shade trees and ran a construction company. It was here that Milarch learned the nursery business, working long, backbreaking days. "From the time I was eight or nine I worked in the nursery," he told me. "I dug and balled trees, dug holes to plant them in frozen ground, hoed, and did all kinds of things like that. My father was a driven man. He had the old-school German philosophy that every day every man earns his salt. He was stern and harsh to the extreme and taught me the discipline of hard work from a young age. There was never any margin. He would probably go to jail today for how I was raised as a child and how I was punished."

Milarch vowed that things would be different in his own family. He is a firm believer in what he calls "mind power," and he tried to instill in his sons the power of positive thinking. "I'd have the kids say these things over and over again: 'The impossible just takes longer.' 'If you think you can't, you're right.' 'Our enemies hold our greatest messages and opportunities.' 'What other people think of you is none of your business.' 'Firmly believe in something, or you will fall for anything.' 'The true measure of a man is not how he falls, it's how he gets back up.'

" 'Boys, what are the limits?,' I'd ask them. 'There aren't any,' they were taught to answer back. 'What can you have?' 'Anything!' Starting when they were two years old, I recorded these sayings on cassette and played them over and over again, overdubbing all the negative things in the world. Hearing positive self-talk is valuable for kids."

In the summer of 1992, several months after he had his near-death experience and his vision, David and his twelve-year-old son Jared gathered the first big tree DNA. Jared had been moved by his father's vision. "When that idea came to him he woke me, and it was like a lightning bolt," Jared Milarch says. "I said 'Yeah, we have to do this.' He always had ideas, and we were always looking for something that had passion in it, but he'd never come up with something like this."

Kerry Milarch was not that surprised by the turn of events. "It was like 'Okay, whatever,' " she says. "He's never been conventional in any way. He seemed to really believe it, but I also thought, 'Maybe he's off his rocker.' "

The younger Milarch thought cloning the champions was good for another reason. "We noticed that the trees on the farm that used to thrive twenty or thirty years ago just weren't growing as well. We thought a lot of the problem was because of warmer temperatures, but some of it was poor plant stock. The shade tree industry has grown the same trees for fifty years, and has not done much improvement in the genetics. When my dad told me about his idea of cloning the champions I thought it might help us grow a better shade tree." For the first few years they sent off cuttings from wild sugar maple saplings. Milarch's father had dug thousands of young trees in the woods on his property over the years to sell at his nursery business, and of those he found some that were extremely fast-growing and cold-hardy from surviving Michigan winters. If they could clone and reproduce those trees, they thought, they could sell a line of Michigan sugar maples. They called the two lines Chippewa Fire and Camp Fire, and they sold them for a while, but more important, the experience taught them how to clone trees.

They collected champion genetics here and there, but it wasn't until the summer of 1994 that they made their first major foray. A bankruptcy had left them with very little, and they had to borrow a pickup truck, pruner, and aluminum ladder from Milarch's father. They obtained a list of the state and national champions from Elwood "Woody" Ehrle, a professor of biology at Western Michigan University and keeper of the state register of big trees, and they drove to Antrim County, Michigan, to a house with the national champion green ash in the front yard, one of fifty-one national champions in the state.

The elder Milarch knocked on the door. "It's a beautiful tree you got there," he told the owner, and he asked permission to take a few cuttings. Jared climbed up the low-hanging limbs with the pruner and brought back a baggie full of new growth snipped from the very end of the branches. Then they took photos of themselves standing by the giant. "If it works," Milarch told the homeowner as they left, "we'll send you a copy." The Milarchs had bagged their first tree. Over the next two days they collected material from four other trees. They shipped the buds to Schmidt's, a large commercial nursery in Oregon that had volunteered to do the cloning for free if they could keep some of the clones.

The discovery and cloning of the Buckley Elm in 1997 created the first national press for Champion Tree. Evelyn Sika, the postmistress at Milarch's rural post office, knew he was working on champion trees and for months had bent his ear about the big tree in the middle of a cornfield south of Traverse City, Michigan, not far from his shade tree farm. And every time Milarch picked up his mail he would agree to go see it, but he never got around to it. One day, Sika made Milarch promise to visit the tree. "You've got to go see the big tree I played under every day as a girl," she said. "You've got to." It was raining that particular day, so Milarch thought, why not today. As his pickup truck approached the field where the tree stood, he was floored. "From a mile away I could see that thing, and I could see how big it was. My God, the crown on it towered over all the other trees. It looked like King Kong. It was eleven stories high and the trunk was eight feet across, a wall of wood. When I got up to it I was at a loss for words."

Moreover, the tree appeared to be an elm. Nearly all of America's elms had died from Dutch elm disease, and here was a giant living specimen. Milarch sent some leaves and bark to Woody Ehrle. It was indeed an elm, and more than four hundred years old. Ehrle couldn't believe it either. It must have survived, he thought, because of its remoteness.

Milarch had heard stories about the big hardwood forests that had stood here in his great-grandfather's time, and finally seeing one of the biggest trees for himself was like a window into the past. The reigning champion elm, near Louisville, Kansas, had just been blown up with a pipe bomb and killed, by a vandal or terrorist of some kind, and Milarch helped nominate this tree, which became the new national champion.

The Traverse City newspaper, the *Record-Eagle*, ran a front-page story about the giant "lost" elm, centering on the Milarch family and Champion Tree; then the *Chicago Tribune* wrote a front-page story describing it as so big that two ranch houses could fit beneath its crown. The radio broadcaster Paul Harvey mentioned it on his show. The owners put a kitchen chair out in front of the tree with a Tupperware container that held a notebook and asked people who came to see it to write their name and their thoughts. More than a thousand people signed in.

The Buckley Elm died in 2002, of Dutch elm disease, perhaps weakened by the fact that its roots were damaged from plowing in the cornfield around it. Fortunately Milarch and his son Jared, with the help of Dow Botanical Gardens, made three clones before it died, and they planted them in three places, including at the Interlochen Center for the Arts.

When the first stories about Champion Tree came out—in

December 1996, their project was the subject of a cover story in the national nursery trade publication *American Nurseryman*, with color photos spread across five pages—the family started to get occasional unannounced visits from Native Americans. "They call them grandfather trees, and they wanted to know what we were doing," Milarch said. "Because we were messing with their fifteen-thousand-year-old drugstores, and we were messing with the spiritual home of their ancestors, because some Native Americans believe their ancestors live in those trees. But when they heard that the project was helping the trees live on, they were okay with that."

Something that struck Milarch as odd was the fact that no one had thought of cloning these big old trees before. Cloning in itself was not difficult; any nursery could clone a fruit tree. Cloning shade trees was more difficult; only a few nurseries were doing it. No one had ever succeeded in cloning the ancient trees; it was thought to be nearly impossible. But everyone knew that genetics could be critical to a tree's health and longevity. Here he was, a nurseryman with little formal schooling, and some of the country's top tree experts were agreeing with his idea to clone trees to preserve their genetic record. What other simple things weren't known or imagined about trees, he wondered.

In the early 2000s, Kerry Milarch bought a book called *Arboretum America: A Philosophy of the Forest* by a botanist, medical biochemist, and author named Diana Beresford-Kroeger. Beresford-Kroeger's message is that the role of forests in maintaining the health of the natural world is far more extensive, and far more critical, than is generally appreciated. We know that trees filter water and air, and that they provide habitats for all manner of insects, which are the first level of the food chain. What is generally overlooked, though, she argues, is the role of the chemicals emitted by trees in the maintenance of the biosphere. "Kerry handed it to me and said, 'You have to read this book.' I was flabbergasted when I read it," Milarch says. "She was the first person who saw the big picture, the global role that trees play. She knew about the medicine in trees, the aerosols, the critical environmental role they play in the world, and the same mysteries of trees that I knew. I found her number and called her, and we bonded immediately and knew we had the same task." The task, he said, was to reforest the planet.

Beresford-Kroeger had emigrated to Canada from Ireland in 1969. Her research ranges from scanning microscopy on trees to foundation cardiovascular research for heart transplants and artificial blood at the University of Ottawa School of Medicine. For the last thirty years she has conducted research in her 160-acre garden and arboretum, funding the work herself. Her books have attracted the attention of some well-known scientists. E. O. Wilson told me he finds her work fascinating. "Her ideas are a rare, if not entirely new approach to natural history," he said. In a foreword to one of her books he referred to her as part Druid and part scientist and wrote, "Beresford-Kroeger speaks for the trees as well as it has ever been done." Dame Miriam Rothschild, a distinguished naturalist, has raved about Beresford-Kroeger's notion of bioplanning—a sustainability system that includes planting trees and other plants according to where their aerosols will do the most good.

"When you peel an orange and get a cloud of mist in the air, that's an aerosol," Beresford-Kroeger explained to me when

I visited her rural home near Ottawa, Canada. She pointed to a towering wafer ash tree. "It's a chemical factory," she said, and the aerosols the tree emits are part of a sophisticated survival strategy. When the showy, creamy, greenish white flowers bloom in the spring, they broadcast terpene oils into the air, which repel mammals that might feed on the tree. But the ash needs to attract pollinators, and so the flowers also have a powerful lactone fragrance that appeals to large butterflies and honeybees. The chemicals in the flowers in turn provide protection for the butterflies from preying birds, because they have a bitter taste. The trees emit medicinal compounds as well. Fox, deer, and other wildlife rub against trees and shrubs, while birds, bats, and honeybees feed and fly among the leaves, and all of them are showered by a range of antifungal, antibacterial, antiviral, and anticancer compounds that protect them from disease and infection. While a few scientists over the years have done some research into these ideas, Beresford-Kroeger's work has gone a long way to promote a little-studied field.

ON A WARM day in 1970, Dr. A. Kukowka, a professor of medicine in Greiz, Germany, had spent a couple of hours puttering in his garden beneath the crown of four massive yew trees. As he described the incident later in a paper in a German medical journal, he was suddenly overcome by nausea, headache, and dizziness. He felt disoriented and lost all sense of time. Visions of vampires, vipers, and other "diabolical" images played in his mind's eye. His limbs became weak and he broke out in a cold sweat. Then the experience shifted and he felt as if he were under a large dome with angelic music playing. He saw visions of a paradise and felt "indescribably happy."

Dr. Kukowka's doctor found no obvious explanation, such as dehydration, for the bizarre sensations. Kukowka said he was able to replicate the experience beneath his yew trees, but he wrote that subsequent journeys troubled him and he stopped his experimentation. Dr. Kukowka hadn't ingested anything from the yew; he had only stood beneath its crown. The active ingredient that had caused the powerful effect upon him, he presumed, came from a chemical in the tree, likely a terpene, aerosolized by the warmth of a summer day.

Terrestrial, as opposed to atmospheric, aerosols are released from the leaves of trees and other plants, sprayed out of something called *glandular trichomes*—microscopic structures on a leaf that look like the onion domes of Russian cathedrals, thousands of which grow at the end of a slender stalk. These biogenic volatile organic compounds include an array of alcohols, esters, ethers, carbonyls, terpenes, acids, and other compounds, and they have a lifespan of anywhere from minutes to months. The abundance of the chemicals depends on a range of factors, from the amount of sunlight to ambient temperature to the genetics of the tree and the species. Their role in the life of the tree, the forest, the earth's ecosystems, and the atmosphere is poorly understood largely because they are difficult to study and they have been considered unimportant. That's begun to change as microanalytical techniques have improved.

In California's Sierra Nevada, researchers took samples from the air just above a remote forest, away from the contaminating effects of urban areas, and found that the air contained a hundred twenty chemical substances, though only seventy could be identified. In French Guiana, researchers sampled compounds emitted after they damaged leaves and bark in the rain forest, and they were surprised to discover 264 different volatile organic compounds in 55 tropical tree species—an average of 37 per species.

Some aerosols rise into the atmosphere and travel hundreds of miles. Others affect the immediate area of the tree, repelling threats or encouraging useful organisms. Some roots emit a volatile substance that attracts useful fungi. Black walnut trees emit a chemical aerosol called juglone as an aerosol, which repels competitive nearby plants by inhibiting respiration and repels some insects as well. The release of these chemicals varies from hour to hour, day to day, month to month, even year to year. The chemicals act in the environment on their own and also combine with other volatiles from other sources, and shower the earth with their essence.

Isoprene is the most abundant forest aerosol, with forty thousand different chemicals. More than half a billion tons of isoprene are emitted annually by the forests of the world, and one of its key atmospheric roles is to protect the trees from solar radiation—in other words, it acts as a natural sunscreen for the planet. Terpenes are the most abundant isoprenes. The world's evergreen forests emit two trillion pounds of dozens of different terpenes each year. Among other things, terpenes are a natural pesticide.

Studies of the boreal forest, the great shield of forest that stretches across the northern latitudes of North America and Europe, show that the emissions of terpene aerosols from the forest into the atmosphere double the cloud condensation nuclei, the microscopic particles that cause the formation of clouds. Because this forest-generated cloud cover keeps sunlight from reaching the earth's surface, the study's authors argue that one role of forest aerosols emitted across the planet is to act as key regulators of a balanced climate: in the northern latitudes, with the onset of the cold of winter, aerosols from conifers diminish and allow more sunlight to reach and warm the planet; when spring arrives and temperatures warm, the trees emit more of these volatile aerosols, which rise into the atmosphere, create more cloud formation, and moderate the heat.

These aerosols play a micro role as well. How plants interact chemically with other plants has intrigued gardeners and researchers since the time of the Greeks and Romans, who noticed that some plants emitted chemicals that "sickened" the soil. The field of aerosol study was formalized in 1937 by a distinguished Austrian plant researcher, Dr. Hans Molisch, a chemist and botanist who would become president of the University of Vienna. He noticed in some simple lab experiments that ethylene, a gas released by early-ripening apples and pears, induced nearby late-ripening apple varieties to ripen earlier. He coined the term *allelopathy* for this new field of study, which, derived from Greek, means "mutual harm."

In the 1920s and '30s, research on these plant substances became popular, and much of the work centered on one type of aerosol, the *phytoncides*. The name "phytoncide," which means "exterminated by the plant," was coined in 1937 by Boris P. Tokin, a Russian biochemist at Leningrad State University. Tokin found thousands of these substances in the air around trees and other plants and hypothesized that these chemicals helped the plants maintain their health and ward off attackers. He also believed that phytoncides influenced everything from microorganisms to higher order mammals, including humans, scrubbing the air of a range of disease-causing organisms. To this day the role of phytoncides is widely recognized in Europe and Asia, where Tokin's work is well known, but his papers and the work of others were never translated into English.

In the West these substances are known as *allelochemicals*. They were studied as a possible source of medicines in the 1960s, but have since fallen out of favor, although lately there is renewed interest in the chemical ecology of trees. "There are three hundred thousand different plant species," said Dr. Gary Strobel, a veteran plant researcher at Montana State University. "Only a tiny fraction of them have been explored thoroughly to see if they are applicable to human ailments."

In the early 1980s an ecologist at the University of Washington, David Rhoades, proposed that a tree attacked by insects signaled its brethren when it was under siege, telling them to take appropriate countermeasures—the release of a chemical that upsets the digestion of the attacking bugs. Rhoades thought the trees communicated chemically through their root systems, but his colleagues Ian Baldwin and Jack Schultz were able to show that it was actually airborne hormones that warned other trees of the attack. The popular media, including *People* magazine, picked up on the story and wrote of the scientists who studied "talking trees." The popularization caused some colleagues to sharpen their knives. In 1985 a review of the existing science was published in *The American Naturalist* that concluded that there was nothing to the notion of signaling between plants. Rhoades left his career and bought a bed and breakfast.

"That was the end of this field for fifteen years," says Rick Karban, a professor of ecology at the University of California, Davis, one of those who recently picked up where Rhoades and the others left off. Critics were wrong, though, says Karban, for trees and other plants are far savvier than is generally believed, and are much more than sticks of wood with leaves. "Trees have been poorly studied because it's hard to get to the canopy, where a lot of the action takes place, and they're slow-growing. But trees and all plants are up to a lot more than we're aware."

Sophisticated chemical communication is one of the things trees are up to. Karban has studied this phenomenon in sagebrush in the steppes near Truckee, California, and above Mammoth Lake, because sagebrush contains a key hormone at levels hundreds of times higher than that of any other tree. Sagebrush plants recognize themselves as distinct individuals, he says, and moreover can also recognize their cloned relations.

When sagebrush is attacked by grasshoppers, which eat leaves, the plant releases a hormone called methyl jasmonate, which communicates to other parts of itself, and to neighboring plants, that it's time to produce something called *phytoalexin*, which can kill or seriously damage the intestinal tract of a feeding insect. "Other plants respond to these chemical cues and become more resistant, so they don't get eaten," Karban says. "Getting eaten is a drag, so over millions of years those plants that can defend themselves do better."

Methyl jasmonate is a fairly recent discovery, only twenty years old, and the handful of scientists in the field have just begun to tease out some of its properties. "Every year the list of the things it does gets longer," Karban says. "And it's active at ridiculously small concentrations." Studies conducted on plant aerosols to date have detected about fifteen different compounds in sagebrush. Newer technologies have begun to show that there are many more hormones, and that the situation is far more complex than anyone suspected. "We're measuring things in the highest concentrations, which are the easiest to measure, but they are not necessarily key," Karban explains. "Fancier toys can detect far more of these, an order of magnitude more. Moreover the chemical milieu is so important that if you take these substances and isolate them you get bogus results." In other words, the synergy among these regulating chemical mists is vital, yet they have only been studied in isolation.

While the study of chemical interaction between trees and other plants is a very small field, there is virtually no study of how the aerosols from trees interact with humans in the environment. There has been a great deal of study of the chemicals that trees broadcast into the environment, however, and this wide array of compounds is being used or studied for healing. One of the sharp, pungent smells in the air during a walk through a pine forest on a warm day is *pinene*, a monoterpene that has been shown to relieve asthma, perhaps by reducing lung inflammation caused by ozone trapped in the lungs. Another aerosolized monoterpene is *limonene*, which, along with *perillyl alcohol*, a limonene derivative, has been widely studied and has demonstrated a range of positive health effects, among them the ability to dissolve cholesterol and gallstones and to prevent asthma. Most significant, though, is its role in cancer—in animal studies it has a robust impact on cancer at all stages. "As a chemopreventative, limonene is very, very good," Dr. Michael Gould told me. Gould is a professor of oncology at the University of Wisconsin and researches the anticancer properties of limonene and perillyl alcohol. In one study he found that limonene in the food of rats prevented chemically induced breast cancer. "There's no question about its role in preventing cancer in animals, and there is evidence it's a cancer preventative in humans," he said. Perillyl alcohol has also shown anecdotal success in treating breast cancer, though, as Gould told me, "it takes a lot more of a dose to treat cancer than it takes to prevent cancer." It has performed quite well in early stage clinical trials for colorectal cancer, leukemia, and pancreatic cancer. One study showed that limonene fed to rats as part of their diet caused "complete regression in the majority of advanced rat mammary cancer." In another study it was shown to reduce the stress response in rats, which may be one reason people feel good among the trees. Other terpenes have similar properties.

Oral administration of perillyl alcohol didn't perform as well in humans as it did in animals, so Gould and others are looking at alternatives—a cream for skin cancer, for example. Clovis O. Fonseca, a Brazilian medical doctor and professor of neuro-oncology, uses inhalation therapy with perillyl alcohol to treat brain tumors that don't respond to surgery or chemotherapy, and his attempts show great promise.

Researchers in the Far East take seriously the notion that forests hold a chemical secret to health. In Japan, Russia, Korea, and elsewhere in the world the positive effect of forests on human health is widely accepted. *Shinrun-yoku*—"forest bathing" or "wood-air bathing"—is a bona fide field of study aimed at understanding what's at work in the moist, fragrant air of an old-growth forest.

In 2000, for example, researchers at the Nippon Medical School took twelve healthy men, from thirty-five to fifty-six years of age, out of Tokyo and into the forest. For three days they followed a regimen: the first day they walked among the trees for two hours, the second day for four hours, and on day three they offered blood and urine samples and filled out a questionnaire. They were sampled a week and a month after the trips as well, and these results were compared to samples taken after walks on normal working days in Tokyo, in areas without trees.

Analysis from the samples taken after the hikes in the forest showed significant increases in "natural killer," or NK, cells, which prevent the formation of tumors; an increase in anticancer proteins in the cells; and a reduction in the concentration of adrenaline in urine, effects that lasted a week after the trips. Alpha and beta pinene were found in the air in the forest, but not in the city, and the researchers assume phytoncides to be the active ingredient in the health effects. Other studies of people who have spent time among trees have shown lower concentrations of the stress chemical cortisol, lower pulse rates, lower blood pressure, greater parasympathetic nervous system activity, and less sympathetic activity, which means that people are more relaxed.

Methyl jasmonate, the plant hormone in Karban's sagebrush studies, which is present in many other trees and plants, has also been found to be a highly effective chemopreventative and chemotherapy agent and is sold for use as an inhalant in natural medicine circles. Remember Karban's statement: "It's active at ridiculously small concentrations." He was talking about its effect on the plant and insect world. But could it be that it is active at ridiculously small concentrations in humans as well?

The big question about the possible role of using forest aerosols as medicine is dosage. Do these substances enter the lungs and bloodstream in large enough quantities to have an effect? Some scientists say the quantities in the air are likely too low for clinical effects compared to the dosage that comes from a pill. Of course no one has measured how much of the substances is getting into the blood from breathing in the aerosols. But if breathing microscopic amounts of pollutants can give us lung or other cancers, or ingesting the microscopic amounts of chemicals called endocrine disruptors can cause problems for us, isn't it possible that tiny amounts of other chemicals might make or keep us well? As Karban said with regard to hormones as messengers between plants, the chemical milieu—the mix of chemicals that travel together—may have a robust synergistic effect that we aren't aware of. Or there may be a chemical in the human body that activates these substances because we have, as a species, coevolved in close proximity to trees and other plants. This is another subject that falls outside the ring of light, and the way to answer these important questions is to research the subject.

But some of the beneficial effects of trees and vegetation on humans may also arise from simply being around trees, since even the experience of seeing trees through a window seems to have a positive effect on health. Dr. Frances Kuo is one of the associates at the Landscape and Human Health Laboratory at the University of Illinois at Urbana-Champaign. Kuo and her colleagues are also on the trail of why trees and other vegetation make us feel good.

Kuo and her colleague, University of Illinois landscape architecture professor William Sullivan, undertook two studies in Chicago's notorious crime-ridden housing projects, the since demolished Robert Taylor Homes and the Ida B. Wells Housing Project. In the first study, published in 2001, they found that people who lived in housing units with access to, and a view of, natural landscape settings that were otherwise identical to housing units where a natural view was missing had substantially fewer aggressive conflicts with family members. Fourteen percent of residents living in buildings without trees threatened violence against their family with a knife or gun, while just 3 percent did so in buildings surrounded by trees and other greenery. The other study found that the highest crime statistics for residences at Ida B. Wells correlated with places where there was no view of, or access to, nature.

Kuo and her colleagues have also studied the effects of green space on children with attention deficit hyperactivity disorder. In one case they looked at the ability of children to concentrate before and after a walk in the woods, and before and after a walk through urban Chicago. The kids, scored by people who did not know where they had been, were substantially better able to concentrate after a walk in the woods. Another study found that girls in a housing project with a view of trees had better self-discipline—they could concentrate better, inhibit impulsive behaviors, and delay gratification, which meant better grades and better life decisions.

Researchers in Europe and Japan and in other parts of the United States have made similar findings. In a very large study, of 345,000 people, Dutch scientists found the amount and proximity of green space to a person's home was a reliable predictor for generally improved mental and physical health. People who lived within a kilometer of 10 percent green space had anxiety disorders at the rate of 26 per thousand, for example, while for those living within a kilometer of 90 percent green space, the rate of anxiety disorders was 18 per thousand. Depression rates were similarly reduced. Green spaces, the researchers wrote, create a halo of improved health around them.

Kuo believes the evidence for the human need for trees is strong but says the subject needs more study. Personally, however, she has no doubt that the lack of trees is a cause of some of society's biggest ills. "A disappearing urban forest leads to a psychological, physical, and social breakdown," she says. "Just as animals in unfit environments develop certain behavioral and functional pathologies, we may see more child abuse or crime or other problems when people live in unfit environments."

The mechanism for these effects is not known. Some researchers speculate that natural settings such as parks mean more chance for social interactions, or it might be because trees clean the air, so there are fewer pollutants and less stress on the human body. A leading theory, though, is that a human-made environment of objects—cars, buildings, or other structures—requires high-frequency processing in the brain. A landscaped environment doesn't require concentration on one or a series of objects, but allows the observer to relax his or her attention, which then has systemwide effects, from reduced muscle tension to lower heart rate and a generally less stressful physiology.

It may be that there is something else at work altogether, or a combination of things. We can dissect the elements that appeal to us about going to visit our mother, for example: the smell of our childhood home, the look of Mom's face, how relaxed we feel in her presence. Have we really explained love, though? E. O. Wilson coined the term *biophilia hypothesis* to propose that humans have an instinctual and deep emotional relationship with nature that it is part of our subconscious. Love may be what is at work in a forest.

For Diana Beresford-Kroeger, there is no doubt that trees help maintain human health, and the health of the rest of the natural world, as they constantly shower healing chemical mists into the air and over the land and water. "These substances are at the heart of connectivity in nature," she says. "In a walk through old-growth forest, there are thousands, if not millions, of chemicals, and they play an untold number of roles. What trees do chemically in the environment is something we're only beginning to understand. And what's happening in the natural world with this web of chemical relationships is something that we ignore at our peril." It's no wonder that when David Milarch contacted Diana Beresford-Kroeger after reading her work, she took to Milarch's big idea and offered to help him.

CHAPTER 6

Willow

THERE ARE BETWEEN three hundred and four hundred different species of willow, from ground-hugging shrubs like the two-inch-high dwarf willow to the weeping willow, the champion specimen of which is over a hundred feet tall.

The willow is one of the most useful trees in the history of mankind. Tribes across North America used its strong, light wood for crafting things as diverse as baskets, cradles, furniture, fish traps, bows and arrows, sweat lodges, wands, whistles, and travois. More recently the willow has been used to make artificial limbs, brooms, cricket bats, and apple crates.

The medicinal properties of the willow are robust and varied, and can be traced to an abundant chemical produced by the plant called *salicylic acid*, which can be used as a mild antibiotic, analgesic, fever reducer, and anti-inflammatory. The use of the willow as a remedy goes back to the time of the Egyptians, when it was a staple in the indigenous medicine cabinet—papyri describe willow leaves as a cure for fever. In Alaska, in the spring, native tribes eat the tender shoots of willows mixed with seal oil as a source of vitamin C. The Cheyenne used it as a poultice for bleeding cuts, and the Cree used its shredded bark as a sanitary napkin "to heal a woman's insides." Fevers, body aches, headaches, and a range of other problems were all addressed by drinking willow teas, applying willow poultices, or chewing willow sticks.

In 1763, a British vicar and pioneering scientist named Edward Stone, suffering from ague, an illness with symptoms that include fever, fatigue, and pain, walked through a meadow near his home at Chipping Norton in England's Cotswold Hills searching for a possible remedy. Stone believed in the "doctrine of signatures"—the idea that where a plant grows provides a clue to its applications. Because willows were abundant in swamps, he reasoned, they might be a treatment for fever, which also often occurs in swampy regions. He tasted a piece of bark; it was bitter and astringent, like the Peruvian cinchona tree, which is the source of quinine, used to treat malarial fever in South America. Stone gathered and dried out a pound of willow bark to make a powder. For the next five years he tried out the drug on fifty feverish people, and it worked quite well. "It's a powerful astringent and very efficacious in curing agues and intermitting disorders," he wrote.

In 1828, German researchers isolated a bitter yellowish crystal from willow bark, which they named *salicin*. A decade later, French chemists created a synthetic substance based on salicin called salicylic acid. Physicians administered the acid to patients with rheumatic fever, and the symptoms subsided. In 1899, chemists at the drug company Bayer in Germany created another derivative, acetylsalicylic acid, a buffered form of the acid, which had fewer side effects—in particular, it didn't cause gastrointestinal irritation. They called it *aspirin*. Acetylsalicylic acid inhibits the production of fatty acids called *prostaglandins*, which are created during the body's fight against fever and other infection and cause pain when they appear around wounds and in muscles. Prostaglandins also appear in the brain's hypothalamus, which regulates the body's temperature, and those fatty acids trigger the fever. When the world was struck by the flu pandemic in 1918, the new wonder drug proved its mettle. Now eighty million aspirin tablets are taken each day worldwide.

Aspirin remains something of a miracle drug. Beyond head and body aches, taken regularly in small amounts it can prevent heart attacks and strokes. Studies show that aspirin can also play a powerful role in preventing mortality from lung, colon, breast, prostate, rectum, and esophageal cancer. And one study showed that people who use aspirin and other anti-inflammatory medication on a regular basis have a substantially lower risk of Alzheimer's disease. Its only drawback is that it can cause serious gastrointestinal bleeding in some people.

Diana Beresford-Kroeger believes that the compounds from the willow are nature's way of caring for creatures that live in and around the waterways. The different species of willows along a creek are, she writes in her book *Arboretum Borealis: A Lifeline of the Planet*, "united to produce salicylate aerosols as antibiotic, antifungal, and aseptic air cleansers. . . . They are highly water soluble, and are added by atmospheric pressure to solubilize into water surfaces. They dissolve into the chemistry of the fresh water as protectants of fish, waterfowl, and underwater life."

Beresford-Kroeger believes that the lack of the dense blanket of forest and other vegetation that once covered the country is the cause of many seemingly unrelated problems, from the disappearance of honeybees to human illness. In her opinion, there just aren't enough trees still standing to produce these important medicinal molecules.

CHAPTER 7

David's Tale

THERE IS ONLY ONE native coast redwood forest in the world, in a heavy fog belt 30 miles wide and 450 miles long, between the California shore and the Sierra Nevada to the east and from south of Monterey to just across the Oregon border. In 2007, David Milarch called to tell me that he had received a small grant for a new Champion Tree project: cloning the largest redwoods. His plan was to use climbers to take cuttings from a few trees; eventually he wanted to take cuttings from a hundred redwood trees across their range to clone, move, and protect the old-growth genes. The new trees with old-growth genetics would be planted in "living archival libraries," like the other trees, and he also talked of wanting to use the clones to recreate an old-growth redwood forest near Golden Gate Park and to plant them in other countries with hospitable climes to protect the old-growth genetics should something destroy the only population of redwoods in California. In October I flew to San Francisco to meet him there and to write a piece for the *Times* about cloning the redwoods.

It's not unthinkable that a disaster could befall the redwoods. Sudden oak death is a fungal disease that showed up in tanoak trees in 1995, a species that grows in the redwood ecosystem. It has killed more than a million trees. What if a similar disease or pest were to decimate or destroy the redwoods? Redwoods are famously disease resistant, but in fact the fungus *Phytophtora ramorum* has affected them—along with Douglas fir, maple, and other oak species—though it only makes them ill; it doesn't, as a rule, kill them. Could it, though, evolve to become more potent and kill the big trees as the planet warms?

DAVID MET ME at the airport. I hadn't seen him in a few years, and he was a little heavier and his hair a little whiter. He looked every bit, as his son Jake describes him, like a cross between Kenny Rogers and Santa Claus. We drove for two hours up the coast, along the black rock cliffs above a frothy sea to an old-growth redwood grove. When we walked into the cool, dark cathedral of giant redwood trees, we stepped into another world. A dead, decaying redwood, a crumbling tapestry of maroons, wines, and russets, lay on the ground like a beached whale, the top of it well above my head. The living trees soared straight up and out of sight, taller than a football field is long. The odor of damp earth filled my lungs. In some places the dense canopy woven by the crowns of the trees all but blocked out the sun, and a velvety carpet of moss on the forest floor was bathed in a dark emerald half-light. A single brilliant shaft of butter-colored light spilled into a pool on the forest floor, illuminating a clump of sword-shaped ferns, like an epiphany from above.

Every tree can be thought of as an ecosystem, as having a complex, evolving relationship with thousands of other living organisms around them, both above and below ground. Because of its size, a redwood is an incredibly large, diverse ecosystem. Tannin, which is found throughout the redwood—in the bark, the wood, and the cones—repels insects, bacteria, and disease, and volatile oils in the wood keep the tree resistant to rot in this temperate rainforest. Tannin also plays a critical role in slowing the cycling of nutrients in the soil. Branches in the canopy fuse to other trees, which helps keep them stable during storms. There are centuries-old five-hundred-pound wet fern mats in the branches of the trees, more than two hundred feet up, essentially small lakes, where huckleberries, ferns, and even other trees thrive. Salamanders and tiny crustaceans called *copepods* live in these worlds above the world, and researchers think they climb up the bark of the massive trees during rain. The marbled murrelet, a small black-and-white seabird with webbed feet that feeds in the ocean, nests here as well.

As we sat and took in the forest, Milarch had a smoke and began to tell me the long version of his near-death experience. I hadn't heard about it since he'd visited me in Montana. He told me that after his experience, he continued to receive ongoing guidance from a range of sources, from archangels to spirit guides, and that the assistance he receives is largely how he navigates. Sometimes the guidance was about small things—

a storm coming that meant he should think about postponing a trip. I was with him once when he said, "I keep hearing the word 'tires.' " We checked the pressure in the car tires, and it was drastically low, though not visibly so. Sometimes it was about big things. He would hear someone's name and have a strong sense he had to meet them, that somehow they would help with the tree project. He had become psychic, he said, and could sometimes intuit what people were feeling. It wasn't like a radio you could turn on—the messages weren't always clear, they didn't always come when he was looking for help, and they weren't always right. On balance, though, they were very useful—to him and to others. While he was waiting in a veterinarian's office near his home one day he noticed a young woman waiting there as well, whom he described as having a dark halo around her that only he could see. "She was about twenty-five," he said. "I don't usually say anything when I see something like that. You're not supposed to intervene. But I saw how pretty she was, and how happy, and I walked up and said, 'You don't know me, but I have to talk to you.' " After explaining his near-death experience, he continued, "I know your boyfriend likes to drink, and I know you are going to a wedding. Please don't ride to the wedding with him." A month later the phone rang and it was the young woman. She had tracked him down to tell him that she and her boyfriend had broken up the day of the wedding. He had gone alone, had gotten drunk, and was killed when his car hit a guardrail. She wept as she told David the story.

The advice and knowing touched various parts of his life, but the tree project was always the main focus of the help he received. Still, the help didn't mean that he would always meet with success. In 2001, for example, Milarch traveled to California to scout for big trees and stopped at Mission Ranch, part of Clint Eastwood's empire in Carmel, to see the champion blue-gum eucalyptus tree that stands on the property and that has since been dethroned. Eastwood had posed on the cover of *American Forests* magazine next to the huge tree, and Milarch had a sense that Eastwood, who is one of his heroes, would like the project. He mentioned to the property manager his desire to clone the tree. Clint Eastwood himself called back and gave his permission. The next time David returned to California, Eastwood met Milarch and his son Jared, and the actor told them how much he admired the project. The Milarchs took cuttings and sent them to a nursery for propagation. A few weeks later the nursery called—all of the blue-gum clones had died. Eastwood's tree was never successfully copied. Even with guidance, projects still sometimes failed.

I didn't dismiss Milarch's claims about intuition and guidance. In the 1990s, I'd written a piece about George McMullen, a remote viewer, or psychic, who worked for many years with Dr. Norman Emerson, one of Canada's leading anthropologists, to find archaeological sites for excavation. Dr. Emerson was an eminent anthropologist, a professor at the University of Toronto, and a founding vice president of the Canadian Archaeological Association. In a paper he presented at a conference to his peers, he described how McMullen's abilities had helped him locate new Iroquois village sites at a level of detail that would have been impossible without psychic abilities. For my article, I'd interviewed a dozen or so police detectives and archaeologists, and they had all insisted that these skills were very real; they all agreed that while psychics weren't necessarily always right, when they were right, it was often to such a deep and detailed degree that it defied guesswork. So when Milarch described his process of locating donors or figuring out the next step in the Champion project, I was more intrigued than skeptical.

Milarch's beliefs are all experientially based; he had no interest in anything remotely spiritual before his encounter with death. Though as a child he attended a Lutheran grade school for eight years, his spirituality has no relationship to any church. His beliefs in karma and reincarnation, for example, come from an intuitive feeling that he has lived before, not from a reading of texts. The conversations I had with friends and family who knew him before and after his near-death experience confirmed his otherworldly journey as transformative. Yet the more I learned about him, the more unlikely Milarch's path to a modern-day mystic seemed—imagine Ken Kesey's *One Flew Over the Cuckoo's Nest* protagonist Randle Patrick McMurphy as Johnny Appleseed.

Milarch had a difficult childhood working on the family farm. And yet the hard labor had its benefits. "When you work sunup to sundown on a farm, you get strong hands and arms," he told me. "By the time I was twelve I had grown from five foot nine to six foot three. When you are in a man's world, doing man's work, and have a slave driver for a dad and you are the son and have to try and outdo everyone, you have a man's body at an early age. If you put on the harness of hard physical labor, and stay in the harness, it has its rewards."

With Popeye arms and oak-strong hands, wrists, and fingers,

Milarch found that in junior high school his strength was vastly superior to that of the other kids, most of whom were just coming into their bodies. He took full advantage. "Hand wrestling was big back then. In ninth grade there were very few kids, if any, I couldn't rip in hand wrestling, where you lock fingers and bend them back. I not only bent 'em back, I'd put 'em down, and it was painful. They started crawling right up the wall." Not fond of academics, Milarch found his calling more in physical contests. "As kids got older and tougher, and more mean, and there were tougher kids from other schools, I used to break fingers and things," he said. "Not intentionally. But it was nose to nose, and they'd break mine if I wasn't careful."

He created what he calls an "arm wrestling machine" out of boards and old inner tubes and spent hours in the garage building up his arms with resistance training, pulling back an old shovel handle attached to a heavy black rubber strap. In high school, he started drinking and hanging out in bars, arm wrestling for beer money. "It was something I was really good at, and you enjoy anything that you're good at, whether it's chess or crossword puzzles or spelling. I just had a predisposition for arm wrestling and bare-knuckle fighting and things along those lines."

Milarch and a few friends started a gang called the Blatz Gang, named after beer brewed by the Valentin Blatz Brewing Company. Members had tattoos and a uniform of sorts—black T-shirts, black Levis, steel-toed boots, and a chrome chain bracelet welded shut, which they couldn't remove. "We were kids with bad attitudes, a lot of anger, emotional problems, things to prove. We'd pull in the parking lot at a drive-in restaurant and our gang would take on other gangs from other parts of Detroit. It was acceptable then. The Jets and the Sharks and James Dean and all that stuff." Things are different now, he said, and he sounded almost nostalgic for that era, with its own kind of dark chivalry. "Now it's no longer a contest of strength or fighting ability, it's who has the most guns."

Bare-knuckle fighting was also part of the gang wars, and with his strong hands and arms, he was the one the Blatz Gang put up against other comers. "It was usually for money, but sometimes we'd do it for fun at drive-ins or after school. Then there started being real fiscal wagers. Their buddies and your buddies would lay money on the ground with a stone on it and then I would start fighting someone from their side, and whoever lasted the longest and got over there to pick it up was the winner. It wasn't pretty, even if you won it wasn't pretty. Let's just say there weren't a lot of referees around." He retired from gang warfare when he was eighteen, in 1968. "I got shot at, stabbed in the shoulder, and beat up bad with a rumble chain all in one year, and I realized there was not much of a future in it," he said. His sister Kathy was a nurse in the ER in Redford, Michigan, and she alone stitched him up eleven times over the course of his fighting career.

After high school he did a year of college at Ferris State but dropped out when he realized it wasn't for him. He took off and hitchhiked around the country, partaking freely in the 1960s and 1970s carnival of excess. "I arm-wrestled my way from Key West with shrimp fishermen all the way to San Francisco, across the country, wrestling cowboys and loggers and all. People who do this aren't always the nicest people you meet, but I had fun doing it." He laughed, started to say something, and thought better of it. "Let's just say I had a lot of colorful experiences and met a lot of colorful people." With his long wavy red hair, his leather jacket, and a penchant for mind-altering substances, he fit in well with the crowd in Haight-Ashbury, where he stayed for a while. He saw his first grove of old-growth redwoods on this trip and was deeply impressed. But he was dismayed to find that the old-growth giants were still being logged. "They were cutting them down as fast as they could," he said.

In 1970, when Milarch was twenty-one, he drove his chocolate-brown 650 Triumph north from Detroit to the Milarch family farm at Copemish, near Traverse City. His father and others in the family had once lived there, but now the place was empty. For three months he camped out beneath the oaks on the nearby shore of Lake Michigan, near the Interlochen Center for the Arts. Smitten by the beauty of Traverse City, he never went back to Detroit. He dug out a basement on the family property, laid boards over it, and lived there for a while. In the summer of 1976 he met the future Kerry Milarch, who worked at Interlochen. They met at a bar called the Fireplace Inn and their rapport was instant. Two weeks later they sold their belongings and hitch-hiked across the country, not returning for a year. They married in 1977, and not long after that David and Kerry built a home on the family land, on a beautiful spot on a hilltop overlooking a broad valley. They built it themselves with lumber from logs that David's father said he could have if he logged them from the swamp on family land. To reach them, they had to wait until the water in the swamp was frozen, so in January, in below-zero weather, they cut down the trees. A friend with a team of horses pulled them out and loaded them on a wagon. Kerry, pregnant at the time, drove the truck to haul out the wagon. The logs were sent to a sawmill and cut into lumber.

While family life brought out a softer side of him, Milarch still loved drinking and arm wrestling—and collecting good stories, which he has a gift for telling. For instance, he told me of traveling to Whitewater, Wisconsin, in 1980 for a family wedding, where Kerry's uncle asked him if he wanted to head down to the saloon for a couple of beers. "Sure," he said. "Always up for that."

"There was a woman at the bar," Milarch recalled. "She was five eight, weighed maybe two hundred sixty pounds. She was two axe handles and a wedge wide. She had on a pair of coveralls and looked every bit like a Wisconsin farm woman, a good-sized one." Kerry's uncle and the rest of the family challenged Milarch to arm-wrestle her. Milarch scoffed, "I don't arm-wrestle women." After more beers were drunk they called him a chicken and made *bawk, bawk, bawk* sounds, until finally Milarch relented. He sat down across from the woman and they locked hands. Someone slapped the counter, the signal to start the contest. "Well, it wasn't three seconds and my arm was on that counter and she damn near threw me off the chair. This couldn't have happened, I thought. Did my arm fail me or am I dreaming, or have I drunk too much? I just got beat by a woman! A farm woman!

"They said, 'Come on, two out of three.' I said to myself,

'Okay, I better get serious.' I really dug in and I was giving her everything I had and I got beat again! I was almost laughed out of that bar." He was laughing hard, shaking his head as he told me the story.

On the way back to the house, his in-laws revealed their little joke: the ample "farm woman" was the national women's arm wrestling champ, who lived in Whitewater. His uncle laughed so hard, tears ran down his cheeks. "They set me up and it was probably one of the most perfect humblings, which I needed and deserved," Milarch said. "If I knew who she was I'd send her a thank-you card for helping to end all of that ego and silliness. It was a gift, it really was."

The Milarchs were back-to-the-landers, living on a 150-acre farm in a small house with kerosene lamps and a woodstove for heat, raising goats and corn and other crops. David helped start a natural food co-op in Traverse City called Oryana. And there were always the trees. He started growing a shade tree farmer's bread and butter: Norway maples, crimson king, sunburst maples, locusts, and birches. His parents also returned to the farm in the late 1970s. In 1979 his first son, Jared, was born, and in 1981 his second son, Jake. The shade tree farm has always been a struggle, and the family bartered honey and firewood with the doctor for the cost of the deliveries.

In 1982 he had his last arm wrestle with a young man from the hockey team at a nearby college. Milarch beat him handily, to the point where the young man fell off his chair. The man was ashamed that he had capitulated so easily. A week later the hockey player died in a hunting accident, and Milarch was crushed that he had humiliated him so soon before his tragic death. Kerry asked him to quit his pastime, and he has not arm-wrestled since.

Though family was important, Milarch couldn't set aside the bottle. "Drinking was my hobby, pastime, and sport," he said. In June of 1991, when he was forty-three, he disappeared for a few days on a bender. When he reappeared it was at one of his kids' T-ball games. He stumbled out of the car, tripped at the top of a small hill, and rolled down to the first base line in the middle of the game. "The kids thought, 'Oh shit, Dad's here,' " Milarch recalled. After that incident, Milarch told his boys he would quit drinking for good. "I told the kids I would never embarrass them again."

He asked his family to leave him alone for a few days while he quit, cold turkey. He placed a bottle of vodka and a six-pack of beer outside of the bedroom door, closed it, and shut himself in. That way, instead of hiding from the bottle, he said, he would beat it. "There is only one of two ways I'm coming out of this room," he told Kerry. "Dead or sober." After three days with no food or alcohol, with only water, his kidneys and liver started to fail, probably due to cirrhosis. He became extremely weak and sick. A drinking buddy, Larry Roundtree, came by to see him. When Milarch let him into the bedroom, Roundtree knew something was very wrong. He carried David, who was in great pain and having trouble breathing, to his car and drove him to the emergency room in Frankfort, Michigan. Doctors discovered that fluid that his kidneys could no longer process was filling the lower thoracic cavity and pressing against his lungs and his heart. They performed a thoracentesis—the withdrawal of fluid from his chest—and the attending physician told Milarch he would need to go on dialysis.

"Dialysis? I don't think I am going on dialysis," Milarch said.

"You've got to go on dialysis," the doctor responded, "or you won't make it twenty-four hours."

But Milarch could breathe again, and he had made up his mind. Roundtree helped him to the car and took him home to bed. His family returned home to care for him.

The next night, as his wife and mother sat at his bedside, Milarch remembers taking a sip of water and then experiencing a strange sensation. "I knew I was dying," he said, "and I was frightened and full of sorrow, but I was so sick I didn't care. I remember lifting up out of my body. It was so painful and smelled so bad, I was glad to be out of it. I remember thinking I was sorrowful to miss the boys, they were only in grade school. Then an angel came alongside me and said, 'Don't be afraid, we know you're afraid, but we're with you.'

"They were with me as I went through the tunnel of white light. Not pure white—it had a pink and blue ribbon, like a DNA helix. Everything went so fast, it was like getting shot out of a rocket. But the angel stayed with me, and I'm glad because I was really afraid. I started to decelerate, and then I arrived on the other side. I stepped out onto a vista. It was like the most beautiful sunrise you ever saw times a hundred, and there was beautiful music. But the best thing was unconditional love. I felt like when you are handed your firstborn child and hold it for the first time—it was like that times a hundred. There were people in spirit form I thought I knew, and light beings, and it was really overwhelming." If this be death, he thought, it was okay.

Then he described a large male angel with a voice like thunder, who seemed to be the boss of the other angels. The angel approached him and told him he couldn't stay; he had to return to his body.

"I said 'Whoa, wait a minute, wait a minute. Why?'

" 'You have work to do.'

"So they sent me back. And you really don't want to leave that situation, let me tell you. But I had to return to the tunnel of light, and *whoosh,* a thousand miles an hour again. When I returned I remember hovering over the bedroom ceiling and looking down at my body. After renal failure for four or five days it looked and smelled pretty painful. But *bang,* back I went. When I got back to my body, I sat up. It scared the hell out of my mom and Kerry, who were sure I had died. I haven't had a desire to drink since then."

For six weeks he lay in bed, unable to walk. The nerves in his feet had died and they had turned a dark color, a necrosis. He finally went to the doctor, who suggested his feet be amputated. "I said, 'No, that doesn't sound like a good idea,' " Milarch recalled. "They hurt like hell, and they don't look too good, but I've grown attached to them. I think they're staying on." Instead, he soaked them in five-gallon buckets of ice water every day to numb the pain. Now, two decades later, his feet still hurt. "To this day that pain is a reminder," he said. "When I'm traveling and people are having cocktails, sometimes for a minute I think it sounds like a good idea. But my feet are there to remind me that maybe it isn't such a good idea."

After three months of recovery he could walk again, though he was frail and had lost weight, dropping from 220 to 159 pounds.

"I looked like a newborn robin that had fallen out of the nest."
Weak as he was, every day he'd go to the local saloon—and have
a soda pop, "because I said I am not going through life running from this thing, from wanting a drink. For two or three
months, I walked into town and had a pop and stared the urge
down. I looked the devil in the eye, all those bottles behind the
bar, and said, 'Not me, not today.' " He started lifting weights,
five pounds in each hand at first. He placed a sign in his garage
with an affirmation that read, "No Man Should Be Shaped like
a Pear." After a year and a half of workouts he was squatting
hundreds of pounds.

In the early morning hours one winter night in 1992, a few
months after his dying experience, his bedroom lit up, waking
him. "It was like three or four cars were shining their lights in
the window, and it scared the starch out of me," Milarch told me.
"I don't know how Kerry didn't wake up. I put my hands over
my eyes and heard a female voice say, 'Get a pad and pen and
go to your leather chair and write this down.' So I said, 'If you
turn the lights down I'll do whatever you want.' They dimmed
the lights and I got a pad and pen and sat in my recliner. I don't
remember anything after that until six, when I got up to wake
the kids for school. And there on the pad were ten pages of an
outline. I'd never written an outline in my life. I hadn't written
too much at all. I guess that's how music comes through when
people write music. That's how this project was born." When
Milarch showed his wife the pages she was dumbfounded. "You
didn't write this," she said, though it looked like his printing.
"You couldn't have written this, there aren't any spelling mis-
takes and you can't spell."

"I'm pretty sure what I saw on the other side three months earlier was an archangel," Milarch said. "And when they call on you, they need something done, and you're gonna do it because they ask with positive attention and love. You don't forget it right away, either." Again, a belly laugh. "That's what I know."

WHEN I LOOKED into the subject of near-death experiences, I was surprised to find a large body of science and writings about a phenomenon I had never heard of. The first written record of an NDE was authored by Plato, who transcribed the story of a soldier named Er who died in battle and returned to consciousness when the bodies of the dead were collected ten days later. He told a tale of ascending to a celestial realm where he saw a brilliant rainbow shaft in the meadows of heaven and watched as other souls chose a new life in which to be reborn. Cheyenne chief Black Elk, psychologist Carl Jung, pilot Eddie Rickenbacker, and actors Peter Sellers and Elizabeth Taylor are a few who have said their lives were deeply transformed by an experience of dying and then returning. "It seemed to me I was high up in space," Jung wrote of his transformative experience. "Far below I saw the globe of the earth bathed in gloriously blue light. . . . The sight of the earth from this height was one of the most glorious I have ever seen."

A 1992 Gallup poll estimates that eight to thirteen million people in the United States have had NDEs—the feeling they have died, left their body, and embarked, as pure consciousness, on a journey to another place—though Dr. Kenneth Ring, a professor emeritus of psychology at the University of Connecticut, believes the figure is smaller, about three million. Like David Milarch, many who have had the experience say it has brought on a deep and long-lasting spiritual transformation. And the experiences happen similarly in all cultures, and across all religions, even among atheists.

While some skeptics dismiss these experiences as sophisticated hallucinations, possibly the products of a neurobiological system convulsed by the shock of death, there are a number of researchers around the world who take NDEs as a journey out of the body seriously. Though Ring retired in 1996 and has not been involved in the field of NDEs since 2000, I found his work interesting because he had worked on the subjective side of the NDE phenomenon, which was germane to Milarch's experience. Ring's research is based on self-reporting; he has interviewed nearly a thousand people who have claimed to have had near-death experiences and has studied the reports of hundreds more gathered by other researchers. When we talked, I expressed some concern about venturing, as a science writer, into the realm of reports of heaven and angels. He laughed. "I wouldn't worry about it," he said. He told me that near-death experiences are almost a mainstream topic these days among serious scientists. Ring cited Dr. Oliver Sacks, whose book *Musicophilia* begins with an account of a physician's NDE, and Dr. Pim van Lommel, the respected Dutch cardiologist and author of the book *Consciousness Beyond Life,* who has studied the phenomenon and whose groundbreaking thirteen-year study of NDEs was published in the scientific journal *The Lancet.*

I described to Ring the NDE that Milarch experienced, his messianic conversion to environmentalism, and his mission to

clone the biggest trees in the world ahead of a planetary catastrophe. Ring called it a classic. "One of the common thematic
aspects of an NDE is a heightened ecological sensitivity," he
said. "NDEers often come back with a sense of mission, and
it's not uncommon for these missions to be, or sound, grandiose. There's a feeling among many NDEers that the planet is in
peril and unless people move en masse toward a solution and
do something about it, we're all going to go down." One example is Dannion Brinkley, who, Ring explained, was told on the
"other side" that his mission was to help the terminally ill die
in a more compassionate way by providing hospice volunteers.
After his experience, Brinkley set up the Twilight Brigade, an
international nonprofit with thousands of volunteers that offers
hospice care to dying military veterans.

In his books and papers Ring describes a realm beyond this
life where the light beings Milarch speaks of dwell, something
he refers to as an *imaginal realm,* a term coined by the French
scholar and mystic Henry Corbin. "It's a distinct realm, not
physical, but quasiphysical, imaginal, but not imaginary, not at
all," Ring said. "These light beings are nonphysical intelligences
that can communicate information to human beings in deeply
altered states—in NDEs or other states of extremis." Some
believe that this realm is the source of the visions experienced
by everyone from saints to shamans to mystics and poets.

Though Milarch says he encountered angels on his journey, Ring doesn't think of them that way. "I don't know how
to describe these otherworldly intelligences. I wouldn't use the
term angel, I just don't feel comfortable with that because it
has religious connotations that are best to avoid. 'Light beings'

is a better term and is used fairly commonly by people who've had NDEs." In Ring's opinion these beings are very real and are working for the good of the planet. "They are higher intelligences that have an interest in the earth and are worried about the future," he said. "Since they cannot directly intervene in human affairs, all they can do is try to influence sensitive and gifted souls and get them to act on their behalf. It's almost like they are spiritual custodians or stewards of the planet, and they need hired help. People like Milarch are recruits. They are in service to an ecological imperative." Milarch describes it as a "noninterference policy." The beings must allow humans to exercise free will, but they can try to influence some who will choose the right path.

While NDEers are transformed and become more spiritually oriented and compassionate after their experience, Ring emphasized that they are still human beings, with real-world motives and desires in addition to their spiritual agendas. They are so possessed with their vision that they can at times be manipulative in order to accomplish their goal. "They read people really well, but they don't always work with passionate selflessness," he said. "They can exploit people," he adds, though they view it as working for the greater good. I asked Ring about the fact that even though Milarch has had this enlightened experience, he sometimes does things that could be described as unenlightened. It's common, he said. "They usually return from their NDE enlightened in some ways, but with their ego intact"—the ego they had before their experience. "Milarch is certainly not unique in that respect.

"As far as the nature spirits, or *devas*"—a Sanskrit word that literally means "intelligent agent" but commonly refers to a nature spirit—"many NDEers report sensing the same," Ring told me. These spirits are widely perceived by indigenous cultures, by Buddhists, Native Americans, other sensitive souls, and even by some scientists who have struggled to incorporate them into the model. From this frame of reference there is nothing preposterous about what Milarch is saying. It's just that our culture, and our normal science, do not operate out of this paradigm. As a result, people like Milarch are often regarded as flakes. They're not flakes, though—they're on to something.

"We do indeed live in a magical universe," Ring said.

All of this doesn't necessarily mean that cloning champion trees is somehow fated, he explained. "I have known a fair smattering of NDEers who have dreamt these big dreams and have had grand plans that make sense from the point of view of an NDE—the vision they were shown—and nothing comes from it. To have a sense of mission, to act on it, does not necessarily mean it will be fulfilled."

Is David Milarch's story true? After much doubt, and after talking with others who have had near-death experiences, I came to believe him. And I believe his claim to be in touch and guided by his higher self, which Carl Jung describes as a guiding intelligence, and perhaps by other spirit beings. I understood, though, that I didn't need to believe that Milarch's deus ex machina is real. I have not found any scientist who says that protecting the genetics of the world's biggest trees in advance of a possible catastrophe on a rapidly warming planet is not a good idea. In fact, with the recent large-scale earthquakes, hurricanes, and tsunamis we've seen, and world leaders who have abdicated their responsibility to deal with climate change, it seems smarter every day.

Milarch, for his part, remains moved by his experience; now, many years after it occurred, it continues to fire his optimism. "When you have a project and you find yourself in a crunch situation and you absolutely have to get it done and it's overwhelming, and you have half enough help and half enough money, your best bet is to find a farmer," he told me, "because those are the circumstances most farmers operate under. It's the situation Champion Tree found itself in from the very beginning. We went from a family project, to a local, regional, state, to a national project, and now we have several countries asking us to come over and do their trees. We don't have near enough money, but we will have enough. I know it. These trees need to be archived, the sooner the better, and the universe wants it, so we'll get it done. The impossible just takes a little longer."

CHAPTER 8
Redwood

CALIFORNIA WAS SPANISH territory in the 1830s when an illegal immigrant from England named Bill Smith jumped ship and walked ashore, disappearing into the fog-shrouded forest of virgin redwoods. He was joined by two brothers, and together they became the first commercial redwood loggers, probably using a whipsaw mill, in which a large tree was laid on a wooden framework above a pit. One man would stand on top of the log and the other in the pit below, and in tandem they would work a two-man saw to form rough planks. It was the beginning of a massive industry that would cut the heart out of the world's greatest forest.

Lumber from old-growth redwoods is among the best in the world. It has rich, beautiful colors, is lightweight yet strong, is straight-grained, and is resistant to rot, termites, warping, water, and fire. Its most valuable attribute is something called *dimensional stability*—the ability of wood to retain its shape over time. It holds nails fast, for example, a hundred years after they were pounded in, even in a wet coastal environment.

Dismantling a towering redwood with nineteenth-century tools required innovation and gumption; loggers had never dealt with trees that size. The wood at the bottom of old-growth redwood trees is extremely dense, forming a strong base to keep the rest of the massive tree upright. Instead of attempting the impossible task of cutting through the base, tree fallers would cut through softer wood ten to fifteen feet above the ground. They did this by driving a board with a metal cleat on one end into the tree. On this diving-board-like platform, which bounced up and down as they worked, they swung their oversized axes and pulled and pushed enormous two-man saws. If a tree fell too hard, it could shatter and become worthless, and so loggers often made a bed of small trees and branches to cushion the fall. Or the giants were dropped uphill to create the shortest fall path. It could take a week for a crew to fell a single tree.

Waste was rampant. Once dropped, the tree was sawn where its diameter was small enough to get a saw through, as high as two hundred feet above the stump. The rest was often left, or loggers drilled holes, packed them with black powder, and blew the trunks apart, destroying much of the wood in the process. Bull, ox, horse, and mule teams pulled pieces of logs out of the forest on log skid roads. Eventually the huge logs were carried out on special wide-gauge railways built into the woods. Trees were peeled and cut into lengths and shipped out on the rail cars; some logs were so big that only a single one would fit in a car. Single-log trucks were also built to carry giants.

In the early days some redwoods, such as those near San Francisco, were close to harbors and relatively easy to load onto freighters. Other parts of the West Coast forest, though, were rugged with no harbor access. Timber companies designed a complex system of chutes and cables on scaffolding that shuttled wood from the hills out to small specially built schooners that could maneuver in small coves called "dog holes" and pick up the logs.

The early cutting of California redwoods was accelerated by the U.S. Timber and Stone Act of 1878, which allowed federal lands with old-growth redwood forests to be sold to private companies at the bargain rate of $2.50 an acre, with timber value per acre at the time at $1,300. The low price was meant to help small-scale loggers, but most of the land, and with it the timber, went to the corporate syndicates. It was a giant transfer of public wealth to private companies. Still, the logging would go on for a long time before the primeval forest of giant redwoods came close to disappearing. In 1854 there were just nine sawmills on California's North Coast; toward the end of the century there were four hundred. In spite of nearly a century of unbridled, unregulated cutting, only a third of California's unique old-growth redwood forest had been felled.

After World War II, with a housing boom to accommodate returning GIs, a soaring California economy, and improved technology, logging efforts were redoubled. In 1953, a billion feet of redwood lumber were produced—triple the yield of any other year prior to 1950. That level remained stable until the 1970s, when it began to decline, falling to half a billion by the late 1990s.

Despite the massive size of the redwood forest, just small bits and pieces were protected before they were cut. Some 95 percent of the ancient forest was logged, and of the 5 percent of old-growth forest that remains—about 106,000 acres—around 82 percent is protected. That tiny fraction of the original old growth is split among forty-eight redwood forest preserves in California, ranging from the 295-acre Muir Woods National Monument to the largest contiguous old-growth redwood forest, the ten-thousand-acre Rockefeller Forest in Humboldt Redwoods State Park. The size of these preserves was determined by cost and availability, not by whether they were large enough to be sustainable into the future. There are real questions about whether these isolated pieces of forest will survive, even without climate change.

The real problem lies in the fact that what remain are fragments, which are far less resilient than large-scale, intact old-growth forests. *Resilience* is the ability of a forest to suffer droughts, windstorms, insect outbreaks, diseases, and fire and to self-repair, to keep functioning as a forest. Small patches of forest often can't recover from large-scale natural events, many of which are predicted to intensify as the climate changes.

Because trees have a cooling effect, for example, the temperature in the interior of a large landscape forest is often a few degrees lower than at the edges, so while the margins may die, the rest of the forest can carry on. Small tracts, on the other hand, suffer from something known as the *edge effect*. When a new edge of a forest is created by logging or road building in an adjacent forest, sun and wind are able to penetrate the remaining forest. Exposure brings changes as far as a quarter mile from the new edge, causing the forest and soil to dry out and allowing invasive plants to find their way in. Many small forest fragments are all edge and will likely die as the climate warms.

And of course the issue Milarch has raised about genetics is a big question mark. What does two centuries of selecting out most of the oldest, biggest, and healthiest trees from the population mean to the future of the redwood forest? "If we lose genetic diversity, we make it less likely that species can adapt to change," says Reed Noss, a conservation biologist at the University of Central Florida, who has studied the redwoods, referring to the dangers of eliminating large acreages of old-growth trees. "Some species are evolving quite rapidly in response to climate change, so less diversity means less possibility for adaptation. Genetic variability is the raw material that evolution works from. With many genotypes [variety of genes in an individual] the chances for survival are greater."

The great decline in genetic diversity in the redwoods is the result of their range being reduced so dramatically. The loss of so many big trees also means a marked reduction in diversity. "The big old trees in many cases probably had superior genes to those that never got so big," Noss says, "and were probably less inbred." Trees also become more diverse as they age, through mutations.

Another factor in the survivability of forests is forest migration. Trees move, though slowly. As the climate gets warmer, seeds that spread to the north might do well, while trees to the hotter and drier south will die out. Scientists estimate that eastern hardwoods are migrating north at the rate of sixty-two miles per century. But forest migration to more favorable habitat as the climate changes is thwarted by development and other barriers.

The problem of forest fragmentation is not limited to the redwoods. The great hardwood forests of the East Coast are in even worse shape. When the Europeans landed, the forests were so thick it's often been said that a squirrel could travel from the Atlantic coast to the Mississippi River without touching the ground. Squirrels can't make that journey anymore. Dr. Doug Tallamy, who heads the Center for Managed Ecosystems at the University of Delaware in Newark, lists figures in his book *Bringing Nature Home* that cut to the heart of the problem. Seventy percent of the forests along the Eastern Seaboard are gone. In the lower 48 states, 43,500 square miles are paved with asphalt—an area five times the size of New Jersey—and 62,500 square miles are covered with sterile, nature-free lawns—an area eight times the size of New Jersey. "We've turned fifty-four percent of the lower forty-eight into a suburban matrix"— homes, roads, malls—"and forty-one percent more into various forms of agriculture." And each year sprawl gobbles up two million more acres of wild land—an area the size of Yellowstone National Park. Large-scale forests are extremely rare; most are highly fragmented, shot through with roads, and riddled with exotic species. According to U.S. Forest Service Research, half of the existing eastern forest is subject to edge effects. But more relevant is the fact that most of the edge effect exists where people live—and that's where the ecosystem services that trees offer, such as cooling and cleansing, are most in demand, and will be even more so as temperatures warm.

CHAPTER 9
Cloning Redwoods

THE MORNING AFTER our visit to the redwood grove where I learned more about David Milarch's near-death experience, David Licata, a filmmaker who was documenting Milarch's work, and I drove to a 306-acre Marin County park called Roy's Redwoods, a few miles out of Mill Valley, where Milarch planned to take his first redwood cuttings. It was a cold, damp Bay Area morning, with a low ceiling of fog engulfing the tops of the trees.

James and Thomas Roy were brothers who were given this land in 1877 by Adolph Mailliard as settlement for a $20,000 debt. Mailliard was the grandson of Joseph Bonaparte, the older brother of Emperor Napoleon Bonaparte and king of Spain and Naples. The property changed hands a couple more times, was slated to be developed for a subdivision, but then was purchased in 1978 for a county park. For a while, hippies lived in some of the giant hollowed-out tree trunks. In 1984, George Lucas filmed scenes from *The Ewok Adventure* here. The three redwoods Milarch planned to clone are four and five centuries old and between two and three hundred feet tall, with a circumference of up to thirty-eight feet. They are named Gramma, Twin Stem, and Old Blue. It's a verdant spot, with a large meadow where weddings and other celebrations are staged.

The best chance for successful cloning lies in using the "sun needles," the new growth in the tips of the branches, which are most vital; they reach up to the sun vertically, and sometimes look like small Christmas trees. The first branches on a redwood, though, are often a hundred feet or more above the ground. (Milarch had arranged with Bartlett Tree Experts, one of the nation's largest tree service companies, to send two tree climbers to Roy's Redwoods.)

Climbers use ropes and harnesses to scale these big trees, similar to the equipment used by a rock climber. When the arborists arrived they cordoned off a circle around the bottom of one of the trees with yellow warning tape. Wearing yellow climbing hard hats and harnesses, they used an eight-foot-tall line launcher—a slingshot on the end of a long metal handle—to fire a weighted throw bag tied to fishing line over a branch where the sun needles grow, and then used the small line to haul up a heavier line. Once the climbing rope was in place, the climbers pulled themselves up the single strand, using both their feet and hands. When they reached the branches they sought, they swung out to the end to remove ten-inch-long cuttings with a saw. Some of this action was captured in a "helmet-cam" that one of the climbers wore.

Milarch had some high-caliber scientific help on the trip. He had invited Dr. William Libby, a professor emeritus of forestry and genetics at the University of California, Berkeley, and a board member of the Save the Redwoods League. They hadn't met before, but after Libby talked to Milarch on the phone, he became intrigued with the idea and offered to bring his considerable expertise to support the cloning efforts.

Libby has written four books on genetics and clonal forestry and has published more than two hundred papers on dendrology (the formal name for the study of trees), forest genetics, genetic conservation, clonal forestry, and biodiversity. He has practiced his own assisted migration, planting California redwoods in New Zealand, where he lived for several years, and taking twenty thousand sequoia seeds to Turkey to establish a population of sequoias in the Taurus Mountains there. He has a collection of several unusual redwoods growing in his suburban yard from clones of the trees he has studied, and one grown from a redwood seed that orbited the moon.

"Glad you could come, Bill," David said when we were gathered, putting on the charm. "When I started making calls to do this, someone told me when it comes to redwoods, you're god with a small g."

"Don't feel like you need to use a small g," Libby joked.

What, I asked Libby, as we stood next to Gramma the redwood, did he think of the plan to protect the genetics of the oldest redwoods in a living archive? "He's picked a bunch of interesting trees," said Libby. "He's gotten trees that have done well for a long time, that have done well in various environments. But a whole lot of things go into living longer, not just genetics. No one can be sure these bigger trees are better trees, though some of them likely are." Libby's opinion was similar to that of Frank Gouin, who had cloned the Wye Oak. We can't prove they are better trees, but like a racehorse or a champion show dog, it's common sense that genetics plays a large role in those traits. And protecting these champions for study in case they disappear could well be useful. "Set aside the guys who really got old so we can study them, absolutely, it's a good idea. It's going to be very nice, if we are able to keep civilization going thirty years from now, that scientists can go out to where clones of these trees are growing in accessible archives and study the genetics with techniques we don't even have yet. If we archive them, future scientists won't have to go throughout the range and get individual permission for each tree, which could keep a project of this scale from happening. And cloning gives you a reliability and level of control over the genetics that you don't have with seedlings."

An authenticated clone of a tree is critical to archiving its genetics—that's the idea at the heart of Champion Tree. "You have to archive the known genetics and know what the attributes of the tree are, the known age and other knowns," says Milarch. "You can only do that with a clone. Even if it's a seedling growing under the tree, you don't know. Did a squirrel carry that one in?" But Libby had doubts about whether trees this old, or older, could be cloned. He had cloned redwoods but was unable to get trees much over 120 years to take, presumably because they are so mature. Cloning the champion tree of every species is therefore picking one of the most difficult trees of each species to clone.

"It's really embarrassing how little we know," Libby said, speaking not just about tree genetics but about trees in general, echoing things I had heard from others. "There're lots of questions for a researcher to ask. Why, for example, is a redwood a youngster at four hundred years of age, while a Monterey pine has generally checked out before it's a hundred? One of the things we know the least about is what is going on underground, the microflora and microfauna that live in the soil around the roots, for example. We don't even have good lists of what is living there yet, let alone what functions they have. They are very important and are likely to be more sensitive to climate change than the trees themselves."

Libby was talking about a tree's *rhizosphere*—the vast complex root system and the soil and the microorganisms affecting, and affected by, the roots. It's a large ecosystem unto itself and is far more complex than the tree aboveground. Roots are vital to the tree's search for nutrients, and its many miles of roots reach into every nook and cranny, working symbiotically with soil, microbes, fungi, and fauna such as nematodes, mites, and spiders to bring sustenance to the tree. Roots feed bacteria in the soil with a range of energy-rich exudates derived from photosynthesis by the tree's leaves, for example, and the bacteria, in turn, process nutrients with the help of fungi on the roots. Until the 1990s the realm of the rhizosphere was largely an unexplored frontier, though in the last decade some gains in understanding have been made. There's an international rhizosphere conference each year that draws hundreds of root researchers, who are fond of quoting Leonardo da Vinci's "We know better the mechanics of celestial bodies than the functioning of the soil beneath our feet."

The global tree canopy is a similar story. Researchers have only recently begun to ferret out what is going on in the canopy of trees because it has been so difficult to access. Steve Sillett's work in the redwood tree canopy is a first for the redwood forest, and researchers have been working in other canopies around the world as well. They have discovered that canopies are an astoundingly rich and complex world—the highest biodiversity on the planet lives in the canopy of tropical rainforests, more than thirty million species—and they have barely scratched the surface. It's the part of the forest that feeds wildlife—a "gigantic food factory on stilts," according to one researcher.

ONE OF THE most important reasons for cloning the old redwoods, or any long-lived tree, is that they may well have a memory for survival encoded in their genes. This idea is based on something called *epigenetics*—the word means "over the DNA," since the changes don't affect the DNA itself but its expression— a field that in the last decade has quietly revolutionized the understanding of genetics. In 2000, Randy Jirtle, a professor of radiation oncology at Duke University, performed an experiment with a type of mouse that carried the agouti gene, which makes the mice appear yellow in color and susceptible to cancer and diabetes. Just before conception and throughout their pregnancies, the mice were fed onions and other foods rich in methyl donors, chemicals that prevent the expression of harmful genes.

The pregnant mice didn't give birth to sickly mice—in fact, their offspring were not only more robust, they showed no evidence of their parents' susceptibility to disease, demonstrating that the epigenetic signals that come from the environment can be passed on from generation to generation. Once thought to be an unchanging code containing the instructions for life that was passed from one generation to the next, DNA is now believed to be mutable, affected by adjoining chemical switches that play a regulating role in how those instructions are expressed. Those chemical switches are altered by the experiences in an organism's life. In the case of the mice, it was diet. Other factors that could change the epigenetics include family experience, stress, or environment. These changes are not necessarily passed on through sexual reproduction, but are passed on when trees are cloned.

In the case of a tree, available nutrients, extreme temperatures, drought, or response to a pest or disease could play a part in a tree's epigenetics. It is not overstatement to say that a tree has some type of memory, and this memory may be a big part of what helps the champion trees survive. "We've seen it with insect attacks," said Libby of epigenetic responses. "A tree once attacked by insects is sometimes able to set up some kind of defense against them the next time around. And there's even evidence that they put out some kind of pheromone that helps the trees around them, and those trees become resistant to the insect as well."

There's something similar at work with disease, called *systemic acquired resistance*, a phenomenon in plants roughly equivalent to a mammal's immune system. In the 1980s, California's

Monterey pines, which grow in the southern end of the redwood belt, became afflicted with a disease called pitch canker. Scientists, in an effort to preserve the forest, inoculated some of the trees with the disease to find out which were susceptible to it. The resistant ones were then cloned and the survivors planted. When the trees that survived were inoculated a second and third time, each time they were better able to withstand the disease.

A case can be made that the DNA of old-growth trees contains a library of knowledge vital to survival of the species, and the older the tree and the more traumas it has survived, the bigger the library. The tree has gone to school on pests, climate swings, and diseases, and if those hardships don't kill it, as the saying goes, they make it stronger, and it can pass the benefits of its experience on to its cloned offspring. Are these attributes the key that will get these trees through the gauntlet of climate change? Only time and more research will answer that question. But it's one strong argument for relying on protecting proven survivors for the possibility of afforestation.

Or the answer may lie not in clones of the big trees that get planted, but in the clones of their offspring. Bill Libby, for example, has planted sequoia seeds gathered from a grove in Converse Basin, near Bakersfield, California. These sequoias are only three centuries old, offspring of nearly two-millennia-old parent trees. Libby hypothesizes that the younger trees produce better seeds than seeds from their parents—they grow faster and are much more vigorous, and they could be better suited for planting on a warming planet. Libby explains that the old sequoias came of age and survived stresses from long ago, whereas the younger trees survived pollution and other modern insults from the onset of the industrial age but missed other things. It could be that the younger trees have less "epigenetic loading" and so might respond better to modern stresses.

As we chatted, Libby said he thought there was an important nonbiological reason for cloning the redwoods: their tremendous iconic status. They are arguably the world's most loved tree, and cloning them could bring attention to the plight of the world's forests and the effort to assist the trees in their migration to other parts of California and the world. "There's no other tree that has the reverence that exists for coastal redwoods," said Libby. "They have a powerful presence, you could even say they are magical. I've seen European foresters cry when they've stood at the foot of some of the trees for the first time."

By the end of the day Bartlett's tree climbers had taken dozens of cuttings from the three trees at Roy's Redwoods, dropping pieces of green foliage to the ground as they worked. Milarch and others gathered the pieces of future forest into a plastic ice chest. It was an unusual trip for them, the climbers said. While they had climbed many trees in their careers, they had never climbed a redwood to create a clone.

The sampling from Gramma, Old Blue, and Twin Stem was well covered by the media, a testament to the fame of the redwoods and the novel idea of cloning them. By now Milarch knew the media well and could provide them with the sound bites and the drama they were looking for. "Okay, we made a huge mistake and stole our children and grandchildren's heritage," he tells the handful of assembled reporters and later repeats on film for Licata. "We've sawn down ninety-seven percent of all old-growth redwoods. There's three percent left. Of that three percent left, only ten percent is protected. Ninety percent of it could fall to the axe or chain saw. So three-tenths of one percent is left of a forest that was here for ten thousand years. Now if you were down to three-tenths of one percent of gasoline for your car, or three-tenths of one percent of your life savings, wouldn't it be time to do something?" In light of recent sophisticated imaging work, Libby says Milarch's comment overstates the threat to the redwood forest. But old growth is still rare, and from the Associated Press to the BBC to my story in *The New York Times,* people heard a message about the cloning of the California redwoods.

MILARCH TOOK THE cooler that contained the redwood cuttings and drove them to an office supply store in Mill Valley, where he packed them in a cardboard box and overnighted them to nurseries, including that of Bill Werner, who has long cloned the West Coast champions for Milarch. Werner is passionate about his role as a steward for rare plants, both as a vocation and an avocation. A devout Christian, he often meets with a small men's group to ask for help from the Creator in growing roots on the cuttings from ancient trees.

"It's money in the bank, and it's a savings account our grandchildren will thank us for," Milarch said after he sent off the package and opened a new pack of Marlboro Lights. A few months later, Milarch got word from Werner that there were now two hundred clonal copies of each of the three trees. "Pretty soon we can start reforesting with them," he said. "We'll have old-growth forest trees, and I'd like to put the first one in downtown San Francisco, in Golden Gate Park. It might seem odd, but that was old-growth forest for thousands of years, and it's only been a couple of hundred years that it's been a city. Why not? We can build old-growth forests around the world."

Again, it seemed wildly unrealistic for Milarch to believe he was going to accomplish all this on the budget of Champion Tree. It would take a great deal of funding to collect a random assortment of DNA from big trees across their range—Milarch's hope was to sample a hundred redwoods from their entire 450-mile range—using climbers to take cuttings from branches, along with the cost of travel and shipping, not to mention the cloning and growing of hundreds and thousands of copies. But Milarch had come a lot further, and lasted a lot longer, than I had expected. Just like the old trees, he had learned a lot by surviving this long.

CHAPTER 10

Dawn Redwood

WHILE THE NOTION of moving trees to protect them might sound novel, it has been done before, and in one memorable project, with a relative of the coastal redwoods. In 1949, Dr. Ralph Chaney, a paleobotanist at UC Berkeley, brought a handful of dawn redwood trees to the United States from China and went on to establish a population of the nearly extinct tree across the globe, a move that assured the species's survival.

The coastal redwood, *Sequoia sempervirens,* and the giant sequoia, *Sequoiadendron giganteum,* closely related members of the Taxodiaceae family, are conifers. While redwoods top out at just below four hundred feet and the giant sequoias reach their maximum height at under three hundred feet, the third member of the Taxodiaceae family, the dawn redwood, *Metasequoia glyptostroboides,* has the muscular redwood trunk of the redwood and sequoia but is the shortest of the three, with the height of the tallest trees just over one hundred feet. Unlike the others it is deciduous, and its feathery leaves give it a shimmery, ghostly appearance. When they drop in the fall, the tree is often mistaken for dead.

Like the others, *Metasequoia* was widespread across the world for millions of years, though unlike its relatives it did quite well in cold temperatures. Botanists thought it had long gone extinct; the most recent fossil was millions of years old. So the scientific world was rocked when, in the 1940s, the trees were found in an isolated valley in south central China. They were given to Dr. Hu Hsen-hsu in Beijing, who, in 1944, sent some specimens to a mentor, Dr. E. D. Merrill at Harvard's Arnold Arboretum, who in turn sent them to Chaney, saying that, unless he missed his guess, these were a living version of a *Metasequoia*. Chaney didn't believe it. How could he? The whole world thought they had disappeared millions of years ago. When Chaney received a packet with green samples from the tree, however, which he had seen only in fossil form, he was overcome. Martin Silverman, a *San Francisco Chronicle* reporter present when he opened the envelope, said Chaney fainted at the sight of evidence that the ancient redwood still lived.

Beside himself with excitement, Chaney, in 1948, arranged a Save the Redwoods League–funded expedition to the remote and often hostile Sichuan province in central China, along the Yangtze. He asked Silverman to accompany him. Silverman talked it over with his editor, who liked the idea but said the tree, then known as the Type tree, needed a catchy name. How about "dawn redwood," because it's been around since the dawn of time? he asked.

In China, the men and their party sailed on a steamer up the Yangzte River to the ancient city of Wanxian. Hiking and sometimes carried in a sedan chair, they spent three days covering rough, mountainous terrain. Bandits prowled the region, and on the return trip one of the expedition's guards shot and killed one. Chaney himself very nearly died, though not from bandits or an exotic disease. For three nights he suffered the wheezing and gasping of asthma attacks, because one of his Chinese guards had accidentally stepped on and smashed a small vial of his medication. Still they pushed on to the village of Modaqi. There, from a ridge a half mile away, Chaney saw his first dawn redwood. It was massive—112 feet tall and nearly 8 feet in diameter. It had no needles, and Chaney realized, for the first time, that the dawn redwood was deciduous—that was why it had thrived in extreme cold while the sequoia and coastal redwood did not. Moving closer, the party could see that a shrine had been built at the base of the tree. Chaney was told that the villagers believed a god inhabited the tree, which they called a water fir, and they prayed to it for healing and abundant rice crops.

There were only three dawn redwoods in the village, and Chaney wanted to push on and see the forest of *Metasequoia* he had heard about near a village called Shushiba. He was still laboring to breathe, and Silverman, who was worried about him, suggested turning back. Their guide made the case to keep going. Whether they continued on or returned, he said, Chaney could die. "If the professor sees the forest, and then dies, will he not die with great happiness?" he asked, and Silverman agreed.

Near the remote village the party reached its goal: several thousand dawn redwoods growing in a natural mixed-species forest in a narrow side canyon. Once widespread, these trees, and the few at the shrine, were all that remained of the dawn redwoods. They had survived because of their remoteness.

Chaney and Silverman returned home with thousands of dawn redwood seeds and seedlings. A photo of the balding scientist holding a seedling appeared on the front page of the *Chronicle* with a caption that read BACK TO AMERICA—AFTER A 25,000,000-YEAR ABSENCE.

"Finding a living dawn redwood is at least as remarkable as discovering a living dinosaur," Chaney was quoted as saying proudly. In an early example of what is now called *assisted migration* or *managed relocation*—moving a species to a new home that it can't move to on its own in order to enhance its chance for survival—Chaney sent seeds from the Chinese valley across the world. The dawn redwood now grows in many places, as far north as Alaska, and is a popular ornamental landscape tree.

CHAPTER II

A Light Shines Down

IN 2008, NEARLY eight years after I first met David Milarch, few people were thinking about how climate change might affect them, or whether trees and forests could be prepared for the journey into uncertain times. That fall I went to Traverse City to talk to Milarch again. He wanted to introduce me to a woman he called a real-life angel who would ensure the fate of his project. The day I arrived, amid a swirl of cigarette smoke, a Joe Cocker CD blasting, Milarch drove me around his farm in a pounding summer rain, pointing out the rows of different shade trees, some of them the champion clones. One reason for his persistence with the project, he told me as we drove, is that when he went through the tunnel of light, he was shown, by his spirit helpers, the tough times that lie ahead on the planet—not just hotter and drier conditions, but a landscape of extreme cold, more violent wind, tornadoes, earthquakes, flooding. There was a deep need, he said, to keep the old-growth DNA alive.

"There isn't the science to say this," he reiterated. "Not yet. But I know it's true. We are not smart enough yet to read DNA. I can only imagine what our computing and science and genetic knowledge will be in a hundred years, and I know in my heart that the information in that DNA will be invaluable. The key is that we are preserving a known lineage, unadulterated, all the way back in the history of that tree. We're not guessing, and future generations won't have to guess."

It is impossible to overestimate the deep and abiding passion that many people have for big old trees. Almost as soon as the story about making copies of the giants ran in *American Nurseryman*, Milarch knew Champion Tree had struck a deep, resonant chord. Along with an inpouring of letters and even money, a call came from the governor of Michigan. He and several state officials had read the story and wanted to meet with the Milarchs. David, Jared, and Jim Olsen, a friend and attorney, drove to Lansing, where Milarch thought they might receive praise and help for the project. It was quite the opposite: state officials did not take kindly to Champion Tree. "It was a set-up," Milarch told me. "For four hours they unloaded on us. 'It won't work,' or 'The trees are too old to clone,' or 'A lot of those trees are on state land, and you can't clone those.' And they questioned what we would do with the money from selling trees."

Milarch was crushed, and he expressed his feelings to his son on the long drive back to Copemish. Maybe the project wasn't such a good idea after all, if the governor's office was opposed to it. "Hey Dad, remember all of those things you taught me when

I was a boy?" Jared asked. "Remember the poster of the pelican with the frog in its mouth that said 'Never ever, ever give up,' even if you are like that frog? Here you are, giving up." Milarch appreciated the wisdom of his son's words, and though he had no organization behind him, no lawyers on staff to advise him, no cash, no membership list, no cloning facility, not much of anything, it turned out that he was a fast learner. "The next day we filed to become a nonprofit," he told me. "All of the royalties would go into environmental education for kids."

But as much as David Milarch did to make Champion Tree a success, many more things seemed to happen of their own accord. In December of 1996, he flew to Florida to spend time in a second home the extended Milarch family owned in Venice. One afternoon, as he visited the orchids at the Selby Botanical Garden, the director of the garden stopped him: "You're the Champion Tree guy, the one who's cloning trees, aren't you?" She had recognized him from the story on the cover of *American Nurseryman*. She asked if he would return the following day to talk to their botanist about the project and discuss how Florida might clone some champions. When Milarch arrived, a surprise awaited—four television trucks and other local press. He figured they were there for some visiting dignitary but soon realized they had come to interview him, the man who was going to clone Florida's champions. Speaking before the glaring klieg lights and a swarm of microphones rattled him, and he describes those first interviews as among the hardest things he has ever done.

"They say you have a Florida Champion Tree project," a reporter said. It was the first Milarch had heard of it. "Yes, I do," he responded. It was part of his style—bluff your way to the next level and then figure out how to accomplish the task at hand. During the presentation he agreed to clone two trees, and the story went out across the state the next day, including on the front page of the *Sarasota Herald-Tribune*. "I had no idea how many people read newspapers," he told me. "I never subscribed to a newspaper and I don't watch TV. I'd never heard of a wire service, and I quickly learned what those are." Florida's trees, however, were not cloned successfully. "Both of them did not work," he said. "I did not know what I was doing back then."

On the afternoon of the day the stories ran, the phone at his vacation home rang. A woman told him she had read about him in the local paper and that someone named Bobby Billie wanted to meet him. When Milarch asked who Bobby Billie was, the woman explained that he was the chief and the medicine man of the Seminoles, a Florida Native American tribe, and she asked Milarch if he would meet them at Bok Tower Gardens, a fifty-three-acre historic garden and bell tower atop a mound sacred to the Seminole people.

Milarch knew that trees were very important to Native Americans, and so the next morning he set out on the two-and-a-half hour drive to meet them, stopping thirty miles from Bok Tower Gardens to fill his gas tank. When he went inside to pay, he saw, standing by the counter, a Native American man with waist-length black hair and a white woman, both with big smiles. "David, it's good to see you," the woman said.

"How in the world did you know I was going to be here?" Milarch asked.

"There are a lot of things Bobby knows," she said.

They drove to Bok Tower Gardens in their separate cars. Milarch has broken his back twice and the drive had caused back pain. As they sat beneath an oak tree on a grassy spot near the parking lot, Bobby Billie asked if he wanted help with his sore back. Sure, Milarch said. "Within a minute or two a giant gray squirrel came out of the tree, ran behind me, and rubbed against my back," he recalled. "I stood up and there was no longer any pain in my lower back. Then Bobby Billie said he knew that one of my favorite birds was a red-tailed hawk, and five minutes later a red-tailed hawk circled over us and landed on my arm." Before they left, Bobby thanked Milarch for coming. "I wanted to meet the 'great white bear of the north' my grandfather said would come and work with the grandfather trees," Milarch recounted him as saying. He laughed remembering the incident. "Welcome to the world of Champion Tree." He then mimicked the theme from *The Twilight Zone:* "Doo-doo-doo-doo, doo-doo-doo-doo."

Not long after, when Milarch was back in Michigan, a man named Terry Mock from West Palm Beach, who had also read the newspaper story, called and asked Milarch to return to Palm Beach Island in Florida to clone the national champion green buttonwood tree. Buttonwoods are a tropical tree and have been widely used for landscaping in south Florida. Milarch felt honored and made the trip. Again the media turned out in force. "This time the story of cloning champion trees went national, and I saw, for the first time really, what power this idea had," Milarch said. "It was Jared and me and some friends creating a group with no funding and no infrastructure, but it was going strong. It's always had a life of its own, like a runaway locomotive. And in Florida was the first time I heard the word 'noble,' that it was a 'noble project.' "

Mock, who would become the executive director of the Champion Tree Project, led it to some of its big early breakthroughs. He and Milarch helped clone a different buttonwood, the national champion silver buttonwood, in Key West, none too soon. In 1998 it was nearly killed by Hurricane George, and in 1999 it was mortally wounded by Hurricane Irene. Fortunately the clones took. In 2001, one of the tree's clones was planted in the same spot at the Key West Golf Course where the mother tree had stood for a century.

During the first four years of the project, until 1999, the Milarchs funded the project themselves, largely through Kerry's job as an English teacher at a local college and their meager shade tree income. Milarch had made a fair amount of money for a few years building log cabins, but he lost it all when a worker without insurance was injured and he was forced to declare bankruptcy. "Neighbors chipped in and bought the kids coats and tennis shoes one winter," Kerry Milarch told me. At one point, early on, Milarch decided to look for funding and went to Washington, D.C., to talk to American Forests, a non-profit organization that promotes tree planting and healthy forests. They are the keepers of the National Register of Big Trees, the list of the 826 species that Milarch hoped to clone. They said they weren't interested. Later, as the idea started to gain a startling amount of traction in the media, and perhaps because they viewed Champion Tree as competition to their own project, they went on the attack. They criticized the science and published an article in their magazine arguing against the importance of tree genetics. A "fundamental mistake would be to assume that so-called 'champion' trees are somehow genetically superior to their counterparts. In some cases, in fact, just the opposite is true," wrote the organization's forester, Gary Moll. And in an interview, their executive director told me she believed Milarch had stolen their idea. "He reinvented our program, basically," said Deborah Gangloff, who has since left the organization.

I thought both claims were disingenuous. Milarch and his advisers weren't saying that champion tree genes were superior—just that they might be. And though he had used the public register to look for the champions, no one at American Forests was cloning them. What Milarch had said was true: the champion trees were disappearing until Champion Tree came along. Later, American Forests filed a lawsuit to stop Milarch from using the name Champion Tree. The suit was dropped, but it cost precious resources.

Terry Mock, savvy about navigating the world of nonprofits and high-end donors, brought some valuable connections to Champion Tree that helped to keep the struggling organization alive. In 2000, Milarch and Mock went to Washington, D.C., to plant a champion ash tree at the Marjorie Merriweather estate. General George Cates, a retired Marine general, had been lobbied by Mock, and after the planting he stepped out of the crowd to propose that Champion Tree partner with the National Tree Trust, a foundation created in 1990 by President George H. W. Bush and endowed with $20 million, whose purpose was to educate people on caring for America's urban forests. They in turn funded Champion Tree with $280,000 a year for three years, allowing the organization to carry out its mission. "It was a third of what we needed, but it allowed us to go national," Milarch said, "and to travel more widely looking for trees to clone."

In 2002, Champion Tree was asked by aides to U.S. senator Carl Levin, one of Milarch's home state senators, who had taken a shine to the project, to provide champion red ash clones for a living memorial to the victims of the terrorist attacks on 9/11. On September 10, 2002, Milarch arrived at the Pentagon to plant the trees. Surrounded by batteries of Stinger antiaircraft missiles and vigilant soldiers prepared to respond in the event of another terrorist attack, Milarch stood on the emerald green lawn of the Pentagon and spoke to a knot of misty-eyed people whose relatives had been killed a year earlier. Next to him were Paul Wolfowitz, deputy secretary of defense, and Carl Levin, the ranking Democrat on the Senate Armed Services Committee. Groundskeepers had used a backhoe to dig large holes in the sod, and as the prayers and invocations ended, Milarch and his son Jared pushed the trees—offspring of the cuttings Milarch and his son had collected on their early foray for champion DNA—into the holes and began shoveling moist black earth onto the balled roots. Milarch asked if anyone in the crowd wanted to take part, and dozens of mourning family members waited in 95-degree heat to help with the planting. "It was so sad," Milarch told me. "It was all we could do to stand there without our knees buckling." Milarch remembers a five-year-old girl in a blue dress. Her father was crying but the girl wasn't. "Can I use the shovel?" she asked. "I want to throw some dirt on for my mom." "When I heard that," Milarch said, "I don't know how I remained standing."

When they were done, many in the crowd broke into tears.

"They offer hope for the future," General Cates said of the memorial trees. "We're going through a very trying time in this country right now. We hope these champions that have survived all these other things will give hope to the American people, saying, 'We've seen it all, we've come through it, and we're going to do it again.' "

That same day, at lunch, Milarch learned that the Tree Trust was being dismantled and its funding would be discontinued after that year.

Interest in the Champion Tree Project, fed by high-profile events like these, continued to grow. Through contacts at the National Tree Trust, the Milarchs planted a living "memorial treeway" at the Firemen's Memorial at Calvary Cemetery in Queens, New York, across the river from the twin towers, to commemorate the firemen who had died in the attack. "That was a tough one," said Milarch. "Seeing all those fresh graves with the pictures of the firemen next to them and the moms and dads and wives and kids, that tore me in half." They cloned a collection of thirteen trees, all that remained out of thousands that had been planted under George Washington's direction at Mount Vernon, including white ash and tulip poplars. Milarch then donated fifty copies of each back to Mount Vernon. They also gave a collection of one each of the thirteen clones to Edsel Ford II and Henry Ford, Jr., to plant at the Eleanor and Edsel Ford Estate on the shore of Lake St. Clair in Grosse Point, Michigan. In an Arbor Day ceremony in 2003, Milarch planted a clone of one of Washington's ash trees on the grounds of the U.S. Capitol as dozens of members of Congress watched. They cloned European copper beeches planted by Teddy Roosevelt on his estate in Cove Neck, Long Island, and Thomas Jefferson's oaks at Monticello. The plan is that all of these historic clones will become part of a presidential tree collection called the Mount Rushmore Collection, to be planted someday on the grounds of the U.S. Capitol. Milarch still has to clone elm trees at Abraham Lincoln's home in Springfield to complete the collection.

In 2003, in one of the high points of his career, Jared Milarch and his team set out to clone Methuselah, a bristlecone pine that, at over 4,800 years old, is the oldest measured living tree in the world. John Louth, the manager of the Ancient Bristlecone Pine Forest, had asked Jared and David Milarch to promise not to publish any full frontal pictures of the tree. Fewer than fifty people know its whereabouts, and the U.S. Forest Service, which manages the public land on which it grows in the White Mountains of eastern California, wants to keep it that way, for fear that vandals might damage or kill it. While Jared took cuttings and picked up some pinecones, others stood lookout in case hikers appeared. Louth feared that Milarch's activity might give away the tree's location, and the plan was to scatter and act nonchalant if anyone approached. In another location, meanwhile, Terry Mock took cuttings from the Patriarch, a young bristlecone at 1,500 years old but the national champion because of its size.

The team drove back down the steep, rutted dirt road and overnighted the cuttings to the University of California, Davis, where a propagator attempted to clone them. The clones didn't take, but Bill Werner coaxed seeds that Jared Milarch had gathered. They weren't clones, of course, but because bristlecone pines pollinate themselves, something many trees do, there is a chance that they are 100 percent genetic copies, although no one will know until further study. *The Washington Post* and *The New York Times* both ran feature stories about the attempted cloning of Methuselah.

In 2006, the 3M Corporation, a technology company, called with a request. They had seen the articles about Methuselah and wished to have a copy of a bristlecone pine for the renowned five-hundred-year-old arboretum at Charles University in Prague. Would it be possible to get a seedling? The company's employees held a fund-raiser to raise money to fly David and Kerry to the Czech Republic to deliver it. The Czechs nicknamed the seedling Methuselah Jr., and Milarch was overjoyed. "Einstein went there!" he exclaimed of the university.

THERE ARE FEW things in this world that can compare with the satisfaction of giving a gift of a seedling from a nearly five-thousand-year-old tree to one of the most renowned arboretums in the world. And when Clint Eastwood puts his hand on your shoulder and asks, "Mr. Milarch, how *do* you clone those trees?" you feel you are doing something right. The project was featured in a television special about DNA, and PBS came to interview the Milarchs. David and Jared chatted with Katie Couric on the *Today* show when they did a segment on the Milarchs' efforts to clone George Washington's tulip tree. They were named heroes for the planet by *Biography* magazine.

Still, striking a deep chord is one thing and striking pay dirt is another. It was deeply frustrating for Milarch that while the tree planting idea had brought him in touch with leading figures, it was nevertheless a titanic struggle to find long-term funding to continue the work of Champion Tree. The Tree Trust had been a godsend, but funding was going to run out and the project would be in financial straits again.

Over the years, many things broke Milarch's way. Besides the Tree Trust funding, he received a small grant here, a little money there. But he couldn't, as he said, "drag the big buffalo back into camp" and provide for the project and his family. I watched a number of potential funding sources come his way, show interest, and then disappear. Part of the problem, it has to be said, is Milarch himself; his blunt, working-class style is sometimes a square peg in the round hole of the white-collar world of nonprofit organizations and garden clubs. It was sometimes difficult for this farmer from Detroit—one who had been a bare-knuckle street brawler and a founder of the Blatz Gang, no less—to rein in the rough-hewn, alpha male part of him and move carefully in the cultivated world of high-end donors.

There have been times when it appeared things were going to finally work out, and then, at the last minute, they collapsed in a heap. That was the case with Patagonia, the outdoor clothing company. David and Jared went to Ventura, the company's headquarters, to talk about a line of T-shirts with champion trees on them. Founder Yvon Chouinard toured them around the compound and they discussed the idea of offsetting Chouinard's "paper debt" by planting trees. The company would figure out how many trees had been cut down to make the paper to operate the company and then commit to planting that many new trees. Everything was agreeable. Then the Milarchs and a few of the company officials went to lunch. Afterward David was told that the deal was off. Someone at lunch had been offended by an off-color joke he had told, and she had the authority to call off the arrangement. "To tell you the truth, I don't even remember the joke, but I think it was something about big breasts," said Milarch. "It was just casual. But it constituted the downfall of the deal." He shrugged. "I think she had a chip on her shoulder. I'd probably tell the same joke again."

Milarch often told me that this kind of adversity energized him. "Storms drive the oak's roots deeper," his grandma Duell used to tell him. One might reasonably ask why, if Milarch had the ear of earth spirits and light beings, the task of carrying out the cloning of the champion trees was so obstacle-strewn. "Maybe they tried to help," he said in answer to my question, "but it was human resistance on my part. I am sure that I have been my own biggest enemy every step of the way. I think I needed to learn a lot about being humble. I have more faults and more weaknesses than most people. So part of the process is for me to learn, and they had me work on a lot of my shortcomings."

Still, I was always amazed at the project's resilience. Over and over again, when it seemed as if Champion Tree had run into a blind alley, that the project was on the verge of tipping over and disappearing just like one of the dead champions, David would meet someone who loved what he was doing and who would pony up just enough money to keep things going. Or he would get a call out of the blue—a garden club would invite him to speak, and out would come the sports coat and off he would go to California or Detroit or New York and his mission would be recharged. In 2002, the Garden Club of America gave him its National Garden Clubs of America Distinguished Service Award, and later its Award of Excellence for Conservation and the Garden Clubs of America Zone X Highest Award for Conservation.

In 2008, the clouds finally parted for Champion Tree. Milarch called me with the news that he had found a real-life angel to fund his dream. "It's time to gather the trees to put in the archive," he said. "It's going to be the biggest damn scavenger hunt in the world. We're going to fill up the ark." As he told me about Leslie Lee, a philanthropist and businesswoman who was ready to do all she could to help the project succeed, I was almost as incredulous as when I heard the tale of the roller-coaster out-of-body trip through the tunnel of light. Milarch doesn't always let the facts get in the way of a good story, and I thought that might be happening here. "Leslie said she'll send the jet for you," he said. "You can come here and see for yourself what's going on." The next thing I knew I was headed to the airport, to the fixed-based operations where private jets land. Two pilots greeted me and ushered me onto a small white jet. My first meeting with Leslie Lee, a determined environmentalist, convinced me that in spite of the odds, the grandiose dream to clone and archive not only the country's but many of the world's grandfather trees was going to become a reality.

LESLIE LEE HAD been following the exploits of David Milarch and the Champion Tree Project in the pages of her local newspaper. She had called the project more than once during that time and left messages about the possibility that her giant walnut tree was a champion, but she hadn't heard back. Had he realized who she was, Milarch might have found time to return the calls. In the 1980s, with her then husband Casey Cowell, Lee had helped build a struggling company called U.S. Robotics. Just when computers were taking off, a handful of computer geeks from the University of Chicago developed a new and much improved modem that connected PCs to the rest of the tech world. It made all of its founders extremely wealthy. Lee and Cowell eventually divorced, and she moved to northern Michigan to raise her kids and become a full-time philanthropist.

In August, her landscaper, Russ Clark, a friend of Milarch's, asked him to assess a three-century-old black walnut tree near Eastport, Michigan. Clark told Milarch that he thought it was a champion. "Christ," said Milarch, "I have had ten thousand people calling me to tell me they think they have a champion in their backyard. If I responded to everyone who thought they had a champion that's all I would have time to do." Clark insisted, and Milarch agreed to take a look.

When they arrived, Milarch was greeted by one of the largest black walnut trees he had ever seen, not quite a champion but a behemoth nonetheless. Clark introduced him to Lee, who immediately blurted out, "You don't return phone calls very well! I've been trying to call you for ten years!"

He was taken aback. "I was busy," he said.

"Nobody's that busy!"

"Well, I'm here now," he said. "Let's take a look at this walnut."

The tree was healthy, though in precarious shape because of rotting in some places. Milarch suggested she call a local tree service, who brought in a bucket truck and cabled it with a half-inch steel cable to keep the branches from breaking off the trunk and ground-wired it to protect it from lightning strikes. They took cuttings and cloned the tree.

There was a powerful rapport between Milarch and Lee. Both have a playful sense of humor, and Lee shared David's concern for the fate of the natural world. As he told her about the Champion Tree Project, she fell in love with the vision of cloning and reforesting the world with the genes of champions. The source of the deep connection between them, she believed, was based on their mutual journey to the threshold of death's door and back. Lee had had her own near-death experience many years before, and though the circumstances were different, the effects endured in her as well. "Since that experience I've always been able to go to a place inside and align myself with it, and with what I wanted to do," she told me when we met. "That's a connection I have with David. I love old trees, and with this project there's a promise of excitement and adventure." Clean fresh water is also a funding priority for her—all of the rooms in her mansion on Lake Michigan are named after the world's largest lakes, and she knows how important trees are to keeping lakes clean.

"What's it going to take to do this project?" she asked.

"A million, million and a half," he said.

"Done," she responded. She bought the whole dream, not just cloning and archiving, but also establishing a way to create sustainable forests in cities and towns. Lee established a corporation, though later it would become two nonprofit organizations—Archangel Reforestation and Archangel Ancient

Tree Archive, with offices in Traverse City and Copemish. The latter would gather the genetics of the old trees and archive them, while Archangel Reforestation would use those genetics to replant and reforest around the world. She hired her cofounder as marketing director, and for the first time in his life, Milarch had a job with health insurance and a six-figure salary. Later she hired his son Jared as nursery manager and his son Jake as a climber and propagator. As a gift, Milarch gave Lee a clone of the Hippocrates Tree, an Oriental plane tree beneath which the father of medicine had taught his students. When the tree is seven or eight feet tall in a few years, Lee plans to donate it to the hospital where her father was treated for leukemia, and where, his clinical options exhausted, he allowed doctors to use experimental treatments on him.

The day after I met Lee, Milarch drove me to the Milarch Brothers Nursery. We walked behind a grassy hill, a half mile or so from the road, to a few old-growth white pine and cedar trees missed by the Michigan loggers, who didn't miss very much. They made the other trees around them look scrawny. "All we know is what we experience in our lifetime," Milarch told me. "But what was here for ten thousand years after the glaciers left? When you go to these remnant old-growth areas you see cedars like this, three feet across, and pines five feet across, and you realize you should be seeing these across the whole ecosystem, that these are what was normal for thousands of years. Our natural systems aren't working because what we perceive as normal in our lifetime is way, way off the mark of what should be here. What we have left is the junk of the junk of the junk, after three or four clear-cuts and taking the best of everything. We have genetically inferior trees with genetically inferior immune systems and an inferior filter system for the land. It doesn't matter if the whales are disappearing, or there are elevated carbon levels, or any other environmental problem—trace them back far enough, and the answer is always more trees. But when you go to the tree store or look in the tree catalog, you realize the trees that are there aren't going to save us. The people who sell trees never looked at the genetics of drought resistance, or resistance to warming, and never did the studies to know what trees hold the most carbon."

Milarch had reached a big milestone—he could now stop searching for the money to make his dream a reality and devote his energy to finding the biggest trees in the world, cloning them, growing them, and moving them to living libraries around the world.

CHAPTER 12
Stinking Cedar

THE BARELY EXTANT Florida torreya, or stinking cedar, is the antithesis of the redwood, sorely lacking in anything close to big-tree charisma. Most of the thousand or so individuals that remain look a lot like Charlie Brown's sad little Christmas tree, small and spindly and just a few feet tall. When the seeds rot, they smell to high heaven, like vomit, of all things, an unendearing attribute that earned *Torreya taxifolia* its nickname.

They were once abundant in this part of the world, and as tall as fifty feet. Their fine-grained yellow wood is rot resistant, and the species was decimated by people for fenceposts, Christmas trees, shingles, and firewood. Over the years the few seedlings that sprouted were trampled or eaten by feral pigs, voles, deer, and other wild animals. A fungus took the lives of many. The stinking cedar appears close to blinking out.

The biggest problem of all, though, is that these cedars thrive in cool mountain habitats. During the last ice age they flourished here in Florida, but now it's too warm and they are stuck. In the past, cedars mostly migrated in one way—gopher tortoises would eat the walnut-sized seeds and travel with them in their gut, then deposit them at some distant site more hospitable to both the tree and the tortoise. But the tortoise is now rare here, and the trees can no longer migrate long distances because of man-made barriers.

Perhaps it was the underdog phenomenon, but when Connie Barlow met these puny, smelly, and exceedingly rare little runts, it was love at first sight. In 1999 she stopped along the banks of the Apalachicola River in north Florida at Torreya State Park and came upon the trees. Barlow asked park officials if she could see some torreya seeds, and they brought out two, on a branch, in a jar of preservative. A botanist passionately interested in biodiversity and deeply concerned about its rapid disappearance, Barlow said she looked at the pitiful offering and thought, "Here was the state park named after the tree, and this was the best they could do?"

Months later, unable to forget the struggling tree, she returned to the park and lay down beneath one for a while. "In my magical, mythical way, even though I am an atheist, I asked the tree what it wanted." She divined that the tree wanted a new home in cooler climes before it disappeared. "I made a commitment to that tree that I would do everything in my power to move it." She would become its new animal partner, she decided, and formed a group called Torreya Guardian. But she couldn't move the trees because they had stopped producing seeds. She vowed to find others, and in 2005, she found Woodlanders Nursery in Aiken, South Carolina. Nurserymen there had taken cuttings from the national champion torreya, growing in someone's yard, and another big old torreya, and cloned them. The seedlings came from the offspring.

Though torreyas are *dioecious*—that is, each individual is one sex and needs a sexually complementary tree to reproduce through pollination—it's believed, perhaps in response to the stress of near extinction, that these two old trees had evolved to become *monoecious*—that is, they grew male and female reproductive parts on the same tree, and self-reproduced—"just like the dinosaurs in *Jurassic Park*," said Barlow.

In 2008, *Torreya taxifolia* was rewilded—Barlow and other volunteers planted thirty-one three-foot-high trees in the mountains of North Carolina, in an undisclosed location, fulfilling Barlow's promise. The species *Torreya taxifolia* is on the short list of trees that may have been saved by human-assisted migration, or at least given their best chance for survival.

CHAPTER 13

The Planet's Filters and Other Ecosystem Services

AFTER WE LEFT the farm, David and I drove to the outskirts of Traverse City and stopped in front of a motel, whose name must be withheld to protect the giant black willow tree that grows in front of it. The ridges of the rough bark are as thick as bridge cable, and the trunk is so fat that, were it hollow, several people could stand inside. Milarch walked up to it, grabbed a handful of branch, pulled it down toward us, and showed me the tips, a glistening vibrant green. He clipped the branches, put them in a plastic bag, and placed them in the back of his truck. Milarch believes this is the national champion. "No one knows it because I haven't nominated it," he said. "But I've taken hundreds of cuttings from it and propagated them." It's not the only big black willow he's taken cuttings from; the Michigan state champion willow is also near Traverse City, and he and officials from the Chippewa Indian tribe have taken cuttings from that one, too.

Milarch has a fondness for willows, which grow in abundance here along Lake Michigan. He showed me two willow houses—circles of willow trees planted as whips, or young trees, and tied together at the top; as they grow they become a dense wall of trees and a great fort for kids to hide and hang out in. One of them stands behind the library in Traverse City. In 2008, Milarch and a friend, Hank Bailey, the natural resource director for the Chippewa tribe, cut about a thousand eighteen-inch-long willow pieces, each about the diameter of a pencil. That year, on Earth Day, they took several dozen schoolchildren out to the mouth of the Boardman River, where it empties into Lake Michigan near downtown Traverse City. There they stuck the thousand cuttings into the black soil on the muddy flats. The next Earth Day they planted three thousand cuttings. "The year after that we had requests for ten thousand, and there aren't that many sticks on the tree," Milarch told me. Now Archangel is propagating black willows in a large nursery facility near his home using cuttings from the national champion.

Milarch said that one of the first trees he had ever wanted to clone was the willow, which has one of the largest natural ranges of any tree, growing from north Florida to Canada and as far west as Texas. Milarch felt that every river and stream should have them, since they are an effective way to clean our waterways. "Imagine a beautiful aquarium," he said, "with green plants swaying in sparkling water, and colorful fish swimming around. Now what would happen if someone unplugged the filter? What would it look like a week later? Would the water still be clear or would it be cloudy? Would the fish still be alive, gasping for the last remaining oxygen? What about the plants? That's what we have done by removing all of the willows along streams and lakes and rivers, and all the other trees. We've unplugged the earth's filter system." Willow trees, which are often found on the banks of rivers, may indeed be a way of cleaning up some of this nation's worst toxic waste sites. Buried in the muck along the Boardman are decades of chemical waste, from machine shop oil to cleaning solvent to chemicals used for timber processing, and Milarch hopes that these plantings will begin to remove some of it.

Many people cite the fact that trees create oxygen, the stuff we all need to live. But some researchers say the oxygen created by the world's forests is minuscule compared to the vast amount of oxygen created by the green mass of microscopic creatures called *phytoplankton* that inhabit the world's oceans. Some 20 percent of the earth's atmosphere is oxygen, and 90 percent of that comes from the seas. In cleaning the water and providing essential nutrients for aquatic life, trees also help the phytoplankton thrive. But in addition to cleaning up the water, trees are remarkably good at sweeping large amounts of pollutants out of the atmosphere. American urban forests alone sequester nearly half a billion dollars' worth of carbon and remove air pollutants that would cost nearly $4 billion to clean up in other ways, including some very toxic ones, from lung-cancer-causing particulates to benzene, ozone, sulfur dioxide, nitrogen oxides, and lead—all of them health hazards. Trees also mitigate some obstructive lung diseases. The leading cause of admission to hospitals in New York City for children under fifteen, for example, is asthma, which causes the network of bronchial tubes in the lungs to close up, making it difficult to draw a breath. It's an increasing problem—the rate of asthma among children soared by 50 percent between 1980 and 2000, with rates particularly high in poor neighborhoods. Air pollution exacerbates the disease. Researchers at Columbia University compared admissions to hospitals around the city for asthma and found that admission rates in a given area fell by 25 percent or so when the number of trees there increased by 343 per square kilometer.

Treating water pollution may be the single most critical service that trees offer to the world. One case study is the relationship between New York City and the rolling forests in the Catskill Mountains to the north of the city that form a catchment and filter for the water New Yorkers drink. Concerned that cryptosporidium, a microscopic intestinal parasite, and other waterborne pathogens might find their way into New Yorkers' water supply, in 1989 the federal Environmental Protection Agency ordered the city to build a new water treatment plant at a cost of $8 billion. After much political debate, and the threat of a lawsuit by the environmentalist Robert Kennedy, Jr., officials instead decided that the cheaper and better option was to protect the two-thousand-square-mile forested watershed that naturally filters water flowing into the city, at a cost of about $1.5 billion. The money was spent on such things as buying buffers of natural landscape around reservoirs as filters and crafting agreements with upstate cities and towns to limit development in watershed areas. While the clean water from the forest alone made economic sense, the intact native forest also provides a

host of other ecosystem services, including wildlife habitat, recreation, and carbon dioxide sequestration.

The move to preserve the Catskill forest filter is an exception.
Far more common are the hard lessons learned after forests have
been removed. One of the most egregious examples is that of
the Chesapeake Bay in eastern Maryland, the largest of 130 estu-
aries in the United States. Once teeming with life, it was home
to blue crabs, shad, swarms of anchovies, and schools of striped
bass. "Baltimore lay very near the immense protein factory of
Chesapeake Bay," wrote the *Baltimore Sun* newspaper columnist
H. L. Mencken in 1940. "Out of the bay it ate divinely." This
nearly two-hundred-mile-long, forty-two-square-mile shallow
mix of fresh and salt waters is now a great American ecological
tragedy, a shadow of what it once was. Numbers of blue crab,
the signature seafood of Maryland, have dropped precipitously.
Oyster fishermen once harvested twenty-five million bushels
out of the bay each year; now the take is around two hundred
thousand. Diseases such as mycobacteriosis and Pfiesteria,
from agricultural runoff, are rampant and cause widespread
fish kills. The bay recently experienced its largest dead zone
ever, which stretched for eighty-three miles. Nothing lives in
this oxygen-poor water caused by fertilizer runoff.

Deforestation is one of the largest causes of the myriad problems in the bay. Where fresh water once fell and was filtered
and slowly released back into the bay there are now farm fields,
construction sites, lawns, and parking lots that pour polluted
sediment into the bay. Farms are particularly bad culprits in
this scenario—research shows that river basins with the greatest amount of farmland produce the most sediment, while river

valleys with the most forest cover produce the least. Nitrogen and phosphorus from agricultural fertilizers and poultry waste feed the proliferation of bacteria that consume dissolved oxygen, and the low levels of oxygen create massive dead zones that cannot support aquatic life.

Deforestation along tributaries to the bay means there are no woodlands to hold back the water, and so more water flows faster. That increases erosion, which causes streams to deepen and narrow. Faster flowing streams reduce the number and array of ecosystem services a waterway can provide. The Chesapeake drains thousands of miles of more than 150 major streams and rivers from six states and the District of Columbia, a massive catchment of 64,000 square miles. Broader forested streams flow more slowly than those without streamside forests, allowing more time for contaminants to settle out and be taken up and neutralized by microbes. A study of sixteen tributaries to the rivers that flow into the Chesapeake found that forested streams are significantly better, as much as ten times better, at removing ammonium and nitrogen, and much more effective at breaking down toxic organic pesticides.

Reforestation can help reverse these kinds of problems. In fact, trees could be utilized to remedy a lot of modern-day water pollution problems, including some of the worst kind of human-created chemical waste: dioxin, ammonia, dry cleaning solvents, oil and gas spills, ammunition waste, polyaromatic hydrocarbons, PCBs, and other industrial waste, even the explosive trinitrotoluene, or TNT. The trees take up waterborne toxic waste and neutralize, metabolize, or aerosolize it. *Phytoremediation,* the cleaning up of toxic waste with trees, is a robust

ecotechnology. Instead of digging up waste with heavy equipment, and trucking it to a hazardous waste site, the task can often
be accomplished by planting willows and poplar forests atop the
site. These species are preferred because they grow more roots
than other species; when the trees are buried deep—eight feet
or so—they grow a large, tangled, robust mass of roots, and the
rhizosphere, with its multitudes of fungi and bacteria, is where
most of the magic happens.

Lou Licht is the founder and president of an Iowa-based
company called Ecolotree that installs fields of poplar and willow trees to remediate toxic waste. He uses tough, fast-growing
hybrids, thirsty trees that suck a lot of water out of the ground,
and he plants only male trees, so there is no reproduction and
therefore no little trees to remove. He sets up an irrigation system to water the trees. "Lots of water is necessary," Licht says,
"because water carries the waste around." One willow can process fifteen gallons of waste a day, and a field of a thousand trees
on an acre can treat ten gallons of toxic water per minute.

Just before the water reaches the trees' roots, it hits a zone
around the root as thick as a finger. "The microbial population
is a hundredfold greater here, because it's fed by the root exudates," Licht explains. Exudates are organic sugars and other
compounds, much sought after by the microbes, that precipitate out of the tree through the roots. Licht calls this finger-wide
halo the "root zone reactor," a critical quarter or half inch of
microbial stew that does the lion's share of the work. "The waste
is often broken down right there, or turned into a precipitate.
The microbial jungle under there just rips the stuff apart."

Licht estimates that a willow sewage treatment "plant" costs

25 or 30 percent as much as a conventional sewage treatment facility. It also has lower maintenance and energy costs and is far easier to operate. But not only is phytoremediation far cheaper, the strategic planting of willows and aspen often does things standard methods of toxic cleanup cannot. The natural filters scrub out pollutants that can't be extracted by conventional methods—pharmaceuticals, flame retardants, chemicals from plastic, and other endocrine disruptors. The hardest part of a conventional toxic waste cleanup is often getting rid of the last few percent of waste, and tree roots can often reach and handle that. In 1984 a tanker truck skidded on an icy road near Medford, Oregon, and spilled 111 trichloroethane, a carcinogen used to preserve lumber, which leaked into the groundwater. For years the carcinogen was removed from groundwater by the usual method of pumping out the contaminated water, but the process was only partially successful. In 1997 a phytoremediation firm from Washington State planted eight hundred hybrid poplars, which pulled the rest of the waste out of the groundwater.

Still, man-made phytoremediation forests can't take care of everything. Mercury and plutonium can find their way into leaves and become airborne, causing air pollution, so annual plants that can be harvested and burned in an incinerator, such as sunflowers, are safer on such sites. And the technology can be oversold. If the waste is too deep, or the site too wet, or the water too salty, phytoremediation might not work.

Enköping is a prosperous farming town in central Sweden near the shore of Lake Mälaren. Sewage sludge here used to be dewatered in a conventional sewage treatment plant, but in recent years the city has replaced the plant with a willow field. The wastewater, which contains high levels of nitrogen, is pumped onto about 190 acres of coppiced willows—willows whose main trunks have been cut off, allowing dozens of basal sprouts, or suckers, to grow. This field of dense willow forest takes up and neutralizes the waste. The system treats about eleven tons of nitrogen a year, the production of the entire city, and some phosphorus, which the trees turn into fertilizer. Then the willow coppices are harvested, chipped into small pieces, and used as a biofuel to generate electricity for Enköping.

The Swedish approach, says Licht, is the beginning of the next generation of phytoremediation, something he calls "function-rich"—serving more than a single use. "The question is, how can we make phytoremediation prettier, make it produce food, use it to save endangered species? We already capture greenhouse gases. Can we increase that? How can we grow something with market value, such as fuels or fiber?"

A critical application for strategically placed fields of poplar and willow trees is the scrubbing of urban and rural runoff that washes into rivers, streams, and oceans when rain falls across a landscape. In cities and suburbs, rain falls on pavement, construction sites, and other impervious surfaces that carry oil, lawn chemicals, pet waste, and industrial waste into waterways. Storm water and leaking sewage from aged and broken sewer lines carry viruses, bacteria, and protozoa that cause diseases that can kill and contaminate shellfish, fish, and other sea life. Sea otters, for example, are plagued by a brain parasite that comes from ingesting cat feces. In farm country similar problems are caused by herbicides, pesticides, and fertilizers.

Because this runoff water comes from wide-ranging sources, it's difficult to capture and treat in a sewage treatment plant. "You can't catch all of that storm water and treat it conventionally, it's too expensive," says Licht. "The only way to get streams and oceans cleaned up is to tree-farm our way through it." The water that runs off 80 percent of the soluble soils in farm country is funneled through a very small portion of the total watershed, or drainage, which means that strategically placed willow and poplar farms could be planted in these areas to capture tainted water on which the trees could work their root zone magic.

Along those lines, Licht has just finished a seven-acre willow installation in Seattle to catch urban runoff that flows into Puget Sound. The water Licht is treating flows into a swale and percolates through the roots of a large grove of willows. When a hundred thousand gallons of water are stored, a drain is opened and the water is released into Puget Sound. "Every drop of water passes within an inch of the roots, and the root zone reactor cleans it," says Licht. In farm country, Ecolotree has used a field of trees to reduce 45 milligrams per gallon of the organic waste nitrate nitrogen—more than four times the federal drinking water standard—to less than one milligram. "It's excess fertilizer in water washing off of a cornfield, going through fields, and entering a stream," says Licht. The trees not only neutralize the waste, they use it for their own growth. "The trees love it and grow enormous." So why aren't these organic systems more widely used? "Engineers who design waste treatment are botanically challenged," Licht tells me.

Phyto fields might also be a solution for dying regions of the ocean. In the 1960s there were 49 hypoxic zones in the world's oceans, and that number has doubled every decade since. Dead zones are caused by nitrogen and phosphorus, largely from agricultural fertilizers. As the nutrients run off farm fields and make their way into the ocean, they fertilize the blooms of naturally occurring algae. When the algae naturally die and sink to the bottom of the sea, bacteria feed on them and consume the dissolved oxygen that is needed to sustain fish and other life. As the bacteria continue to eat, the level of dissolved oxygen declines to the point where the ocean in that region is no longer able to support life. The most infamous of dead zones, extending over 8,500 square miles—bigger than the state of New Jersey—is in the Gulf of Mexico, fed by nutrient runoff from large-scale agriculture along the length of the Mississippi River.

PHYTOPLANKTON ARE TOO small to be seen with the naked eye, yet they are vital to all life on the planet. They are abundant in the world's oceans and, through their process of photosynthesis, they are tasked with the job of turning sunlight into food for other sea-dwelling creatures. Experiments have proven that their numbers are greatly enhanced when iron is added to the ocean.

The importance of iron led one path-breaking scientist to make a unique connection. The Erimo Peninsula on the north coast of Japan saw its forests clear-cut and its hills turned into pasture long ago. The change drove off the schools of fish that once teemed there, and caused a decline in oyster populations. Katsuhiko Matsunaga, a Japanese marine chemist, spent years studying the relationship between forests and oceans. His key finding is that even where iron is abundant in parts of the ocean, it is oxygenated, which means it is not readily available for the tiny creatures. What *can* make iron available to phytoplankton to perform photosynthesis, however, is fulvic acid, one of several humic acids that comes from the decay of leaves and other organic matter. The ongoing, natural decomposition of centuries of tree leaves and other material on the forest floor, and the leaking, leaching, and washing of this chemical stew into the ocean, is vital to increasing coastal phytoplankton, and thus the things that eat them, and those that eat them, from oysters all the way to whales. The award-winning chemist spent much of his life working with fishermen in coastal communities to reforest the coast and the banks of rivers and streams to increase fish stocks, and he wrote about it, in Japanese, in one of his papers, "When Forests Disappear the Sea Dies."

These are the understated and poorly researched roles that trees and forests play in maintaining and enhancing the biosphere that earns them the term "ecotechnology." Nothing that the human enterprise does can come anywhere near the elegance and efficiency of a robust global forest.

CHAPTER 14
Ulmo

THE ULMO TREE grows in Chile and Argentina, and is similar to the elm. It is deciduous, elegant in appearance, with a stout trunk, broad green leaves, and beautiful white flowers whose nectar is gathered by bees to make a creamy, highly-prized honey. The dense, hard, decay-resistant wood is also prized for making fences and firewood, and so the tree is in trouble.

There may be yet another reason to value the tree. For the last forty-five years, Gary Strobel, a plant pathologist at Montana State University, has scoured far-flung forests around the world for what he calls the "jewels of the jungle." He doesn't look for compounds in the leaves or the bark of the tree, as most bio-prospectors do; Strobel is looking for fungi with unusual prop-erties in the microscopic crevices between the cells of the new leaf stems. Fungi and other microbes that grow in these spaces are called *endophytes,* a scientific term that means "in the plant." The trees he is examining are those that have grown in the same place for millions of years, which means that the fungus and the tree have had a long time to coevolve. The Ulmo is one of them.

Strobel made headlines in 1987 when he injected elm trees on the Montana State University campus with a genetically engineered bacterium he created from bacteria he found in between the cells of wheat, which he believed would protect the trees from Dutch elm disease. The injections violated EPA rules on the release of genetically engineered organisms, and when a story about the rogue experiment appeared on the front page of *The New York Times,* Strobel was required to cut down all of the inoculated trees.

Some of the compounds Strobel has found in trees have led to patents, including for several different kinds of antibiotics as well as a new type of immunosuppressant drug that is gentler than the ones currently used. He recently found an antibiotic in Bolivia that kills drug-resistant staph, E. coli, and salmonella but is safe to use as a mouthwash. But his most surprising discoveries to date are the properties found in a fungus he named *Gliocladium roseum.* This fungus gives off eight compounds associated with diesel fuel, including octane and heptanes, yet it does not contain some of the polluting compounds found in diesel such as naphthalene. Through a microscope in his campus laboratory he showed me the fungus, which looks like a red, long-stemmed flower. He calls the compounds "survival microbes"; he believes they help the trees ward off disease and insects in exchange for a place to live.

While any possible fuel production from the fungus is still

a long way off, the fact that such an organism even exists is the real discovery. "There is no other creature on the planet we know that can create diesel without refining," he told me. Fungi are known to emit hydrocarbons, but they need processing. This fungus is different—it emits diesel compounds into the atmosphere around it. Pure diesel in its natural state does not need refining from crude oil, and Strobel's fungi can be grown in vats on wood chips or crop waste and produce what the scientist calls "mycodiesel." "The main value of the discovery may turn out to be the genes controlling the production of hydrocarbons and the potential for being able to genetically engineer them into other microbes, such as yeast, which grows faster and could carry out the process more efficiently," Strobel notes.

Strobel's work is yet another argument for archiving rare trees whose properties we do not yet understand. While fungi are commonly researched for their beneficial properties, the fungi that grow in the spaces between the cells of leaf stems are especially interesting because they compete in an environment with far fewer microbes, and less competition means they develop more unusual properties to live compatibly with their host plant. The diesel emissions are likely a kind of natural antibiotic that helps the tree ward off infections. According to Strobel, "There's an untold plethora of unidentified species out there, and they are useful products. Not just for medical uses or for fuel, but for a whole range of things." Perhaps the most precious jewels of the jungle are yet to found.

CHAPTER 15

Ireland

ONCE LESLIE LEE agreed to create the Ancient Forest Archive, the resources were available to launch the project on the scale Milarch had always hoped for. He and Lee decided that the first country outside the United States they would try to clone in would be Ireland. The native Irish old-growth forests, almost completely mowed down and never replanted, are in even worse shape than those in the United States. Only about 1 percent of Ireland's native woodlands remains, the least of any European country. The forest fragments are so small and vulnerable that protecting them as soon as possible is critical.

With polar ice rapidly melting as a result of climate change, and an increasing amount of freshwater pouring into the oceans, much of northern Europe could be at serious risk. Experts say it's very possible that the Gulf Stream, the conveyor belt of warm water that flows from the south and keeps northern Europe warm, could become so diluted with heavier freshwater that it will sink into the ocean. That could plunge Europe into a deep freeze, which could in turn mean widespread forest death.

Or, as is already occurring in the United States, more, or more virulent, diseases could wipe out stressed forests. In 2010, officials from Britain's leading tree groups, Woodland Trust and Woodland Heritage, sounded the alarm about the discovery of a new disease called acute oak decline that has infected both of the United Kingdom's two species of native oak, the sessile and pedunculate oaks. The fungus causes a canker on the tree, which oozes fluid—something known as "bleeding." Then it causes the dieback of the branches, starting at the tips, which finally leads to death, within three to five years. "This is our most iconic tree, it totally dominates our landscape, and here we have a potentially new disease that leads to a rapid decline of the tree," says Andy Sharkey, the Woodland Trust's head of woodland management. Acute oak decline also kills beech, larch, and ash.

One of the first things David Milarch and Leslie Lee did after they formed their partnership was to bring Diana Beresford-Kroeger for a visit to Michigan, first as a consultant and later as senior science adviser. Bill Libby was also hired as a consultant. One of Beresford-Kroeger's first tasks was to choose the first one hundred trees from the global forest to be cloned, the most important trees to have archived in case of a catastrophe. The crew also made a trip to Bill Libby's home in Orinda, California, to talk about the ethics of collecting trees and the ethics of assisting trees in a migration to other parts of the world.

They decided that species would be grown and planted in suitable places where they are native, or moved to places where they had already proved to be uninvasive. Redwoods and sequoias, for example, grow around the world with almost no invasive properties. Meanwhile, the risk of importing disease on trees would not be a problem if cloning was done in certified nurseries, since trees that are shipped between countries from these facilities must have bare roots and be certified as disease free. In the absence of proof that some tree genetics are better than others, proven survivors are arguably the best bet. "If you want to plant a tree and walk away," Milarch said, "where will you get your genetics? From trees that have survived." Moreover, it's not just old trees but old-growth forest that may be critical, and plantings will include native soil, soil microorganisms, and understory species—as much of the forest as possible.

David Milarch, Leslie Lee, Diana Beresford-Kroeger, and her husband, Chris, would go to Ireland for two weeks in early December to begin the European collection, and I was invited to join them. The morning after our arrival, we were met by our driver, Dermot Buckley, who guided us through the small villages and rolling farm fields in the countryside of southwest Ireland. December in Ireland is a damp, dark affair, with nickel-gray clouds hanging low and a sun that comes up late and goes down early. The heart of a North Atlantic winter, though, is the time to take cuttings from the trees, when the buds are dormant and the branches will survive being cut and transported.

Climax oak forests covered more than two-thirds of the country until the sixteenth century, and as a result Ireland was one of the great forest civilizations of Europe. The worship of trees goes back thousands of years in Ireland, to the Druids, the high priests of the Celts, who taught that sacred springs, stones, and groves were a way to connect with the divine. Brehon law, an ancient Irish code of conduct, was centered on the protection of the forest. Each of the original twenty letters of the Ogham, the alphabet of the native Irish, was named after a tree or another plant of the native forest. Trees were sources of food and medicine and inspiration, and were often where teachers assembled their students for lectures. The forest and the native culture were so inseparable that in the early sixth century, Caesarius of Arles, a Catholic leader seeking to convert Ireland, forbade tree worship and urged the faithful to burn up and chop down the "sacrilegious" trees to their very roots. Another Catholic leader, Martin of Tours, found out just how sacred trees were. A crowd of worshippers stood by as he demolished a pagan shrine but mobbed him when he tried to cut down a sacred tree. Centuries later, after the British subjugated Ireland and banned the old ways, the Irish formed outdoor "hedgerow schools" to teach children about their native history, language, and faith. So critical were the forests to the Irish way of life that one of the planks in the Sinn Féin platform was the replanting of the native forests.

In the 1500s, when the Tudors arrived from England to conquer, they started to mine the vast oak woods, caring nothing for proper forestry. "Trees are an excrescence provided by nature for the payment of debts," said the landowner Sir Jonah Barrington in the late eighteenth century, characterizing the prevailing attitude toward the great Irish deforestation. Large trees were taken to be used for ship and house beams, branches were used for barrel staves, which the English used to ship their goods around the world, and the remainder of the tree was used for charcoal to fire iron and glass furnaces. There was also a side benefit for the English—stripping the country bare deprived the Irish rebels of their hiding places.

Today, native trees in Ireland are usually solitary, growing in the middle of a farm field or in a town. The only forests most visitors see are the commercial, monoculture stands of Sitka spruce, a conifer native to North America, planted by Coilite, the Irish forest agency. These exotic forests are planted solely for timber, not for the ecological benefits of a forest or for recreation.

OUR FIRST STOP to collect old-growth DNA was Mount Congreve, a carefully tended English-style garden overlooking the River Suir, where the trees are three and four hundred years old. Mount Congreve is the epitome of the English garden, from the time when Britain's ships ruled the world and brought species to the United Kingdom from all over. The seventy-acre property was owned by Ambrose Congreve, who married into a family of industrialists, the Glasgows, who have cared for these gardens since the 1700s. Congreve, who was 102 when we visited and living in the Bahamas, had given us permission to take cuttings.* Even in the damp gloom, the garden was an idyllic landscape. Diana Beresford-Kroeger clipped several samples of new growth from the branches of two yews that had been planted by the Glasgow family in the 1700s. All of Ireland's yew trees

---

*Ambrose Congreve died in May 2011, at the age of 104.

are descended from just two yews that were brought here from England in the sixteenth century. It's a good thing we came to this garden when we did; there is talk of "developing" this glorious piece of real estate.

As we drove to our next stop we discovered the persistence of the old ways of the forest. On Ireland's most modern motorway, Buckley pointed out a single small, scraggly tree growing in an ocean of grass along the road. "It's the Latoon fairy tree," he announced. In 1996, when the motorway was under construction, a traditional Irish storyteller named Eddie Lenihan called the local press. A hawthorn tree that lay in the path of the road being built, Lenihan claimed, was where the fairies once did battle. A farmer had told Lenihan of driving his cows there one morning and seeing the fairy blood on the ground. Unlike the gentle fairies in Disney movies, Lenihan explained, Irish fairies are fierce, and the blood was likely from battle. "Immediately, he knew that the fairies had been fighting the night before and had dragged back their wounded and dead after the battle," Lenihan told the paper. When construction workers read the story, they refused to cut down the tree. Hawthorn trees are said to have a powerful guardian spirit that becomes angry if the tree is not cared for. At night, however, someone sneaked over to the construction site with a saw and reduced the hawthorn to a stump. Shortly afterward, in what seemed a miracle to the locals, the tree sprouted new leaves and branches. Contractors finally agreed to leave the tree in place.

Ireland is a small country, and by sheer coincidence, Buckley's cousin Andrew St. Ledger was a member of the Woodland League, a group whose motto is "Dedicated to restoring

the relationship between people and their native woodlands."
St. Ledger is also a craftsman who fashions fine furniture and
sculptures out of wood. For years he and his colleagues have
fought the Irish forest policy of planting monoculture timber
farms and have asked the government to begin to restore native
woodland across the country. Buckley says St. Ledger has grown
hundreds of oaks from acorns and has worked on a plan with
others in his group to harness the energy of schoolchildren to
restore Ireland's woodlands. Joining us for lunch at a local pub,
St. Ledger listened as Milarch explained the project to clone the
largest trees, starting with his near-death experience and ending
at the Congreve gardens. "We're here to help Ireland save the
last of its old growth," Milarch told him. "The cavalry just came
over the hill."

Hearing David's words, Andrew became overwhelmed, to
the point where his eyes grew misty. He offered to take us to
some of the old-tree genetics that remain in Ireland. We all
got into Buckley's van and motored off to what Andrew called
the Brian Boru tree. As a fine rain fell, we climbed carefully
over an electric fence and splashed in inch-deep water across a
farm meadow, where a large, lone bull eyed us warily. When we
reached the tree, its crown spread over us, offering shelter from
the steady drizzle. One large branch had grown to the ground,
like a giant elbow, to stabilize the tree. This lone sessile oak
dates to at least the tenth century and is named after Brian Boru,
the last high king of Ireland. The border between fact and fic-
tion is often permeable in Ireland, but the story goes that Boru
rallied his men here, beneath the spreading crown of the tree,
and then headed to Clontarf, where in a bloody battle he and

his men beat the Vikings. That night as Boru prayed in his tent, an enemy penetrated the camp, entered Boru's tent, and ran a sword through him.

Many trees in Ireland have a story to tell.

Despite the rain and slippery bark, St. Ledger climbed the tree and inched his way out onto a muddy branch to take cuttings for us. We had expected that the collection would come from these big old solitary trees, so we were surprised when he then took us to an eighty-acre privately owned fragment of Ireland's forest past called Raheen Woods. A previous owner had logged most of a large tract of forest to farm but had protected this patch as a hunting preserve—a wild wood, as the Irish call it. The noise of whizzing traffic died as we walked into the growth of trees, and the forest hush took over.

Beneath towering moss-coated oaks, on a path carpeted with glistening wet oak leaves, we made our way into a forest cathedral of gnarly trees, a breathtaking remnant of Ireland's forest past. It was a walk through time as well as space, and a treasure trove of ancient forest genetics. Not only are the oak trees here old growth, but the entire ecosystem is as well. Just like oaks in North America, the oaks of the British Isles are a critical source of native biodiversity, home to more moths, butterflies, and other insects than any other tree. More than three hundred species of lichen alone grow on oaks in Britain and Ireland. "There's been oak cover here for eight thousand years," St. Ledger told us. "The forest is a three-dimensional maze of recycling nutrients—passing from plant to animal to bird back to plants again. It's a rare example of the authentic rainforest landscape of Ireland. There's common oak, hazel, holly, alder, willow, mountain ash, cherry, black thorn and white thorn, and an amazing variety of ferns, ground ivy, tree ivy, and brambles. I've seen badgers, pine marten, fox, voles, red squirrels, and deer. It's a symphony of sorts, each playing its part."

St. Ledger was pleased as punch to show us this secret garden, and we were all pleasantly shocked to see something akin to a living dinosaur. We walked around in wonder. Diana Beresford-Kroeger in particular was agog. "I know of no other place like this in Ireland," she said in amazement, "nowhere." St. Ledger had permission from the landowner for gathering, so Milarch and Lee leaned a stepladder up against the trees, took cuttings, put them in baggies, and stored them in an ice chest.

While Ireland's native forests may be Europe's most decimated, native forests across the Continent have similarly been destroyed. They once covered as much as 90 percent of mainland Europe; now they cover about 30 percent and are in fragmented condition. Ireland has the smallest amount of old-growth forest while Finland has the most, with 72 percent. If old-growth genetics are indeed critical to rebuilding forests to survive a changing world, the last chance for Europe is hanging on by a thread in many places. And even these small pieces are not safe. Only a popular uprising in England in 2011 averted the sale of much of the country's public woodlands to private landowners who might well have cut them down.

Toward the end of our stay we drove on to Killarney National Park. We could not take cuttings here but came to look at some of the great old trees of Ireland. We walked along the Lakes of Killarney, past massive trees. But as beautiful as the Killarney woods are, they are mostly parkland and not a wild forest. The park does contain some slivers of old growth, though. One of Ireland's largest old-growth yew forests is found here, its mossy green floor carpeted with red yew berries. We were surprised to see a giant redwood tree lying prone on the ground in the park, planted a century ago and felled recently because of disease.

Killarney includes Muckross Abbey, an ivy-covered fifteenth-century stone structure with a cemetery in front. Major sections of wall and roof are missing, and as we approached, a crew of workmen were toiling, laying new stone floors and rebuilding the two-foot-thick abbey walls. The foreman agreed to show us around, and we climbed construction ladders to the top of the structure. The highlight, when we arrived there, was the view of a giant yew tree planted in the courtyard, its sharp conifer fragrance filling the air atop the church. The yew was planted here in 1448 and is among the oldest trees in Ireland. Churches and monasteries throughout Ireland were often built on sites sacred to the pagans, so that people who came to visit the sacred yew might also visit the church and be converted. Whether the visitor was pagan or Christian, the tree was viewed as a portal to divinity, a place where a seeker could go and have his or her soul lifted to the heavens.

CHAPTER 16
Yew

IN 1974 A YOUNG British man named Allen Meredith went to visit a churchyard in Broadwell, a village in southwest England, to give a talk about bird watching. On a tour before his talk, the guide pointed out a two-hundred-year-old yew tree. As recounted in the book *The Sacred Yew*, Meredith suddenly knew intuitively that the tree was far older. A year or so later he had a dream about a group of elders wearing gowns that brushed the ground and hoods that hid their faces. He was told by the elders to look for the Tree of the Cross, which he understood to mean that the tree was so long-lived that it held secrets to immortality. "I knew right away it was the yew tree," Meredith told the authors of the book. Dream upon similar dream followed, and they brought to mind the strong connection with the yew he had experienced in the churchyard. The dreams spilled into his waking life, and he became obsessed with efforts to find out more about the yew. He pedaled his bicycle around the English countryside visiting the ancient yews to try to figure out their age, and to find out how many listed in old records were still standing. What he found worried him.

Yews grow very slowly, and their wood is very dense. They are nearly impossible to date because, much like a bristlecone pine, they grow in peculiar ways. As part of their aging and self-renewal process, they will sacrifice their heartwood, the denser center of the tree, or other parts, to a wood-digesting fungus and yet keep growing, and as a result, much of a living tree may appear to be dead. Furthermore, the yew's ability to *branch layer*—to grow its lower branches toward the ground and root—can turn a single tree over many centuries into a grove. Because the trees sometimes seemingly stop growing, people think they are dead and cut them down. Meredith discovered that hundreds of ancient yews had mistakenly been cut down this way.

Meredith's dreams and intuitions were insistently telling him that the yews were far older than commonly thought and, furthermore, that they were sacred. Realizing that operating on intuition alone would brand him as a kook and he might never get his message about the true age of the yew trees heard, he set out to school himself in dendrochronology—the aging of trees—and spent months reading in libraries, scouring church records, and writing to experts around the country, even as he continued to visit yews to study and measure them.

In 1980 Meredith met Alan Mitchell, one of the leading tree experts in Britain and the founder of the Tree Register of the

British Isles, which contains records of tens of thousands of Britain's Heritage Trees, the oldest, widest, rarest, and most historic specimens. Meredith told him of his passion for yews and of his visions and showed him his records and research. Deeply skeptical at first, Mitchell eventually came around to Meredith's point of view: that the yews could be thousands of years old. "In fact there appears to be no theoretical end to this tree, no reason for it to die," Mitchell concluded. Many of Britain's tree experts came to believe yew trees were ancient because of this self-taught tree expert. It led to a new appreciation for the tree, and to redoubled efforts to study, catalogue, and protect it. Meredith disappeared from the tree scene many years ago and became a recluse, but the work he began to reappraise the age and importance of the yew is being carried on by others.

Some people believe the yew is the most sacred tree in the world. "The yew is found on every continent in the northern hemisphere and has a spiritual aspect in every country it is found in," writes Fred Hageneder, author of the exhaustive *Yew: A History* and founding member of the Ancient Yew Group, which continues to catalogue and protect the ancient yews of England, some two thousand of which are in their registry. The yew has been worshipped across the world, from Europe to Japan to the Pacific Northwest of the United States. Nature spirits of the Shinto religion of Japan, called *kami*, are said to sometimes take up residency in certain trees, and one of these revered trees, called the Hakusan Jinja yew tree, some forty feet tall and twenty feet around, was declared a god in 1673 because it was believed to harbor *kami*. In the Pacific Northwest the Tlingit tribe used yew wood for death masks and spirit whistles. In the British Isles the oldest yew trees are often next to Christian churches, in graveyards, and in other venerated places such as Muckross Abbey. The yew tree was believed to be a portal to the underworld, and a sprig or branch of yew was often laid in burial shrouds or coffins.

The yew has more earthly attributes as well. With the exception of the bright red *aril*, the fleshy pulp around the seed, the tree is extremely poisonous. A yew concoction was carried as a poison pill by the Celts and others in case they were captured by the enemy. Celtic warriors covered their arrowheads with the sap to increase its lethality. Surprisingly, the yew's healing properties are also more profound than those of any other tree. The Pacific yew in North America provided the natural molecule that has since been synthesized and is, under the brand name Taxol, a powerful chemotherapeutic used to treat breast, prostate, lung, stomach, head, and neck cancers.

The yew has also been praised for its lethal properties as a weapon. The military advantage in the thirteenth to sixteenth centuries lay in longbow proficiency. At the battle of Crécy during the Hundred Years' War, for example, the outnumbered English won, thanks to something called an "arrow-storm"— seven thousand archers firing seventy thousand arrows per minute, "as thick as snow, with a terrible noise, like a tempestuous wind preceding a tempest," wrote one witness. Deeply impressed, the English ordered every able-bodied man to learn to use a longbow, and to own two of them.

The best wood for a longbow was the yew. Bow wood was cut where the sapwood—or outer wood on the tree—and the heartwood meet. Because these two types of wood have

complementary properties, when cut as one piece to form a single bow, a de facto composite, its power was unprecedented. The
heartwood side, which compresses well, faced the archer, while
the sapwood side was best at stretching and lengthening when
the bowstring was pulled. The yew gave the bow a draw weight
of 120 pounds compared with the 100 pounds of lesser wood,
while some had a draw weight of 150 or more, and a range of up
to a thousand feet. Neither solid oak nor armor could stop the
metal-tipped arrows fired from a yew bow. The Gwent archers
of Wales were known for their skill at nailing enemies on horseback to their mount by driving an arrow through the armor, the
leg, the leather saddle, and deep into the horse with the power
of the yew bow. "The strength inherent in a longbow, combined
with the qualities of the 'composite' yew, forged the most effective killing device known to mankind at that time," Hageneder
writes.

By 1470, English yew forests were nearly exhausted, and
the British government realized a need for new sources of yew
wood. They solved that problem in 1472 with a yew tax. Every
ship that unloaded in any English harbor was required to bring
four "bowestaffs" for every tun-tight, or large cask, of wine they
imported. The rush across Europe for centuries-old yew trees
was on. By the early sixteenth century, the yews of Europe were
decimated. To this day they have not recovered, and most of the
yews are solitary trees. One of the largest and most beautiful
exceptions is Kingley Vale, a wildwood yew forest preserve near
the city of Chichester in southern England, with thousands of
trees, some of them up to two thousand years old.

CHAPTER 17

Mystics and Freethinkers

DAVID MILARCH ISN'T alone in arguing for a more expansive view of trees. For fifty years Lawrence Edwards, a geometry teacher who lived in the small, tidy village of Strontian on the west coast of Scotland, went to the park near his home and, through his thick glasses, painstakingly measured the new buds on the oak, beech, ash, elm, birch, and cherry trees there. He meticulously gathered more than a hundred thousand data points, quantifying changes in the shape of the buds, determining how sharpened or flattened they were. Over the decades he found that leaf buds have a subtle yet unmistakable pulse in the fall and winter, growing flatter in shape and then sharpening. His findings were especially unusual because even though trees stop their metabolism in the winter, this pulse continues.

A careful scientist, Edwards, who died in 2004 at the age

of ninety-two, found conclusively that each species of tree had a two-week shape-shifting rhythm geared to the alignment of the earth, the moon, and one of the planets. "When the moon passes through the line of Mars and Earth, all of the oaks do their little dance for a day or two," Graham Calderwood told me. Calderwood worked with Edwards for twenty years and now carries on his research. "When the moon passes through the line of Earth and Venus, the birches do *their* little dance." Edwards saw these kinds of changes in all species of tree he examined except one, a knapweed tree that was situated under the town's only power line. Edwards and Calderwood believe the electromagnetic field created by the power line may block the tree's link with the planets.

Edwards identified another cycle. As the fourteen-day cycle slips a bit in duration over the years, a seven-year cycle, reflected in the alignment of planets, seems to reset the clock, and the fourteen-day cycle becomes fourteen days again. The calculations are sophisticated and precise, says Dr. Jay Kappraff, a mathematics professor at the New Jersey Institute of Technology in Newark, who has studied and written papers about Edwards's work.

But what does it all mean? Edwards didn't advance a theory to explain his findings. A teacher of upper-level mathematics at the Edinburgh Rudolf Steiner School, he was an adherent of the teachings of the Austrian mystic Rudolf Steiner, who claimed to see an etheric field around all living things, a kind of geometric matrix that guides their growth. While the planets and trees would seem to be unrelated, they are, Steiner intuited, connected through this matrix that is invisible to most. Steiner believed that other plants, cells, embryos, even the human heart were connected to this sacred matrix, which could be understood and utilized for great benefit if it were unlocked by a geometric formula.

Edwards's and Calderwood's thorough work is among a great deal of unusual research into trees and forests that hints at how little we know about these beings on yet another level. The physicist Walter Heisenberg wrote that "what we observe is not nature itself, but nature that responds to our method of questioning." As we look into the world of trees, we begin to realize that there are many important questions that haven't been asked by scientists.

Other scholars have noted a link between trees and the cosmos, though it is hard to say what it means. In the mid-1970s, Russian scientists from the Botanical Institute of the Russian Academy of Sciences examined the growth rings of an 807-year-old juniper tree growing on a mountaintop more than 11,000 feet in altitude. The width of the rings showed a substantial slowing of tree growth during the fifteen years following known supernovas—in 1604, 1770, and 1952.

A few years ago, researchers at the University of Edinburgh in Scotland found that the single biggest predictor of tree-ring growth in fifty-three-year-old Sitka spruce trees wasn't precipitation or temperature, as one might expect, but *galactic cosmic rays*, or GCRs. In the years when the most galactic cosmic rays were bombarding the earth, the trees grew fastest, and the connection held up under scrutiny. There is a "consistent and statistically significant relationship between growth of the trees and the flux density of galactic cosmic radiation," the researchers wrote. Galactic cosmic rays are atomic nuclei that have had their electrons stripped away. They originate from outside the solar system, but from within our galaxy. The authors hypothesized about why this might be: the cosmic rays produce nuclei that cause cloud condensation, which diffuses the solar radiation reaching the earth, which may, the authors speculate, increase photosynthesis.

David Milarch, informed, he says, by his imaginal realm sources, also believes trees are connected to the stars. "Trees are solar collectors. Most people equate that with the sun's energy. But the sun is only one star, and there are billions of stars that influence Earth with their radiation—it's called starlight. Trees photosynthesize the energy from the stars just as they do from the sun. I also believe energies inside the earth are transmuted and transmitted into the cosmos by the trees, so the trees are like antennas, senders and receivers of earth energies and stellar energies. I don't know what the energy does, or where it goes, but I think in a hundred years science will be able to measure those energies and quantify them."

Rupert Sheldrake, a Cambridge-trained biologist, plant physiologist, and scientific heretic, hypothesized something similar to the idea of an intelligent matrix in his controversial theory of *morphic field resonance*. In his 1981 book *A New Science of Life*, Dr. Sheldrake proposed that there is an invisible field of intelligent information within and around a living organism—say, a redwood tree. The field contains information that organizes the structure and pattern of activity of the redwood, while the tree in turn feeds information learned from the environment back into the field of information for other, and future generations of, redwoods. The tree's DNA, rather than containing the information itself, is instead more of a bioantenna that sends and receives information to and from the field. In other words, "believing in genes," said Fred Hageneder, a scholar of the mythology, ethnobotany, and mythological aspects of trees who accepts Sheldrake's idea, "is a little like believing the postman created all your mail."

Many ancient cultures have had their own version of an intelligence, or field of intelligence, responsible for different kingdoms on the planet. Buddhists call them *lahs;* in Sanskrit they are called *devas;* to the ancient Greeks they were dryads; Celts and Native Americans also had nature spirits. At Findhorn, the mystical garden in Scotland, communicating with *devas* is a routine part of the farming. "I had never set out to learn to talk with angels, nor had I ever imagined that such contact could be possible or useful," wrote cofounder Dorothy Maclean in her book *To Hear the Angels Sing,* which equates angels with *devas.* "Yet when this communication began to occur, it did so in a way that I could not dispute." Is it possible that Steiner's matrix and Sheldrake's morphic fields are related to *devas* and *lahs?*

When Sheldrake's ideas were published they caused outrage among some in the scientific world. Sir John Maddox, editor of the prestigious magazine *Nature,* called the book "the best candidate for burning" to come along in a while. "Sheldrake is putting forward magic instead of science, and that can be condemned in exactly the language that the Pope used to condemn Galileo, and for the same reason. It is heresy," he said in an interview televised on the BBC in 1994. Sir John's very unscientific condemnation earned him a great deal of criticism, even from people who agreed with him. And it may not have been the best analogy, for in the end Galileo was right.

Few people have written as much about the spirit and speculative science of trees as Fred Hageneder. Born in Hamburg, Germany—another country with a long relationship with trees and forests—Hageneder, fifty, now lives near the scenic harbor town of Swansea in Wales. He has written four books on trees, including his book on the yew. Like many people who have a reverence for trees, Hageneder believes they were vital to the spiritually inclined, not just as objects of veneration but as the source of some kind of native energy. His efforts have been focused on trying to explain that relationship in a way that isn't too far out. "In tree literature we have a huge chasm," he told me. "On one side are ecologists, botanists, and others who wouldn't touch anything spiritual with a beanstick, out of fear that the media as well as academia would ridicule them. On the other side are those neopagan believers, modern druids, or witches who write ever so passionately about spirits and fairies and messages from angels that you sooner or later think, Goodness, get real!" Hageneder tries to bridge the gap and find a middle ground between the two in his work.

He told me about a team in England who studied subtle energy grids across Britain and wrote a book about it, *The Sun and the Serpent*. "They dowsed a few trees in churchyards, and what they found is that each tree is collecting some type of life force from the universe, and as it gathers this energy it appears to spread it out. They found a star shape of lines going out into the land, and a star shape going out from the bottom of the tree. The tree serves as an irrigation system, but not for water, for this life force."

In his book *The Spirit of Trees,* Hageneder gathers together information from a number of sources to show some of the thinking about the myriad roles of trees in the world. How can we presume to know the full role of a forest, he asks, when science doesn't know all of the properties of so basic a substance as water? Water's physical properties are extraordinarily complex and a subject of disagreement among scientists. In the 1990s, Paul Caro, the head of the French National Scientific Research Center, wrote in his book *Water,* "We can measure the varied properties of water over a wide range of temperatures, yet the measurements reveal that water acts like a strange subject that defies logic." Water forms complex and stable clusters, for example, whose structure is more akin to crystals. And there is some research that shows that water, in defiance of the laws of modern physics, has a memory. When chemical substances were placed in water, and the water was then diluted until not a single molecule of them was left, somehow the water maintained the properties of the substances. Many mainstream scientists pooh-pooh this idea because it's outside the known model of the natural world, but the few studies there have been show that something might well be going on. Luc Montagnier, the French virologist who won a Nobel Prize in 2008 for discovering the link between HIV and AIDS, not only thinks the phenomenon is real but has developed a theory. He believes bacterial DNA emits low frequency radio waves that arrange water molecules into nanostructures.

Hageneder also writes about the electrical properties of trees. In the 1920s, a well-respected Yale neuroanatomist, Harold Saxton Burr, who studied the electrodynamics of trees for decades, released his findings: that trees have a steady but fluctuating electrical potential field ranging from 0 to 500 millivolts. In his research, the voltage was at its lowest in the morning and peaked in the afternoon. Burr also noted seasonal variations, with potentials peaking in September on the day of the equinox and hitting their nadir during the spring equinox. He found that light and darkness, the earth's magnetic field, moon and sun cycles, and other factors also impact the electrical field of trees. Five hundred millivolts is a little less than half the voltage of an AA battery. In 2009, researchers at the University of Washington created a device to harness this small, though not insignificant, amount of power. With something called a boost converter, they dialed up the tree's power to create 1.1 volts, enough to run a sensor that can remotely detect forest fires.

Hageneder describes a Czech scientist, Vladimir Rajda, who in 1989 found that the electrical activity of a tree is tied to its biochemical metabolism—in other words, the tree's chemical processes generate an electrical voltage, and the voltage in turn affects the chemistry, in a "soup-and-spark" process similar to the biochemical electrical activity of the human brain. The electrically active part of the tree is the thin membrane of xylem and phloem, layers just under the bark. Rajda also found that the tree's electrical activity is tied to atmospheric electricity, the daily changes of the planet's atmospheric electromagnetic network, and the movement of the sun and moon. Rajda believes that each species is different in its generation of voltage, and that trees can self-regulate this biophysical field.

There's Russian research from the 1970s that shows that light, moisture, nutrition, and photosynthesis are not the only keys to plant growth—an electrical charge in the air also plays an important role. Because the surface of the planet is nega-tively charged and the ionosphere is positively charged, the Rus-sians believe that trees are "continually discharging electrical tension voltage between the earth and the ionosphere."

These electrical properties, Hageneder posits, may make trees a player in the global atmospheric electrical circuit. While the electrical field of lone trees is small, a German physics teacher named Rainer Fischer proposed that the electromagnetic fields of trees are amplified when they are gathered together in large for-ests. As their sap ebbs and flows, so does their electrical charge, and thus their magnetic field. A big question is what sets the planet's electrical currents in motion. "Many theories have been developed to answer this question. But the simplest, the electro-magnetic effect of countless parallel vegetable electrical conduc-tors, has been overlooked completely," Fischer said, referring to the forests. "The strength of Earth's magnetic field is dependent on the density of vegetation. When the vegetation retreats, the magnetic field strength of the earth decreases. Today, exactly this is revealing itself to a high degree." Widespread deforestation, in other words, may have reduced the force of the planet's mag-netic shield against radiation from the sun and cosmos.

There is yet another realm to the study of the hidden lives of trees: some scientists believe that plants can both feel and respond physiologically to human emotion. Many of these ideas were taken up in the 1973 classic *The Secret Life of Plants*. One prominent subject of the book was Cleve Backster, a preeminent expert in lie detection who was an interrogation specialist with the CIA and chairman of the Research and Instrument Committee of the Academy for Scientific Investigation. At age eighty-seven, Backster still runs his own polygraph school, the Backster School of Lie Detection in San Diego. In a 1960s experiment, Backster hooked a polygraph to the leaves of plants. He found that if a leaf was harmed, or even if someone intended to harm it, the polygraph displayed a spike in electrical resistance similar to the response of a suspect who is emotionally distraught when lying. Backster called the phenomenon of plant communication *primary perception.*

Is there more communication going on in the woods than we know? Is there some kind of electrical wave through which plants communicate with one another? An Oregon physicist named O. E. Wagner set up his detection equipment in Oregon's ponderosa pine forest. When he pounded a nail into the trunk of a tree he detected a slow-moving wave that traveled from one tree to others, signaling some kind of distress. He coined the term *W-wave* for the slow wave, which travels at three feet per second. "The tree with the nail put out a tremendous cry of alarm," he told the Associated Press in 1989. "The adjacent trees put out smaller ones." The W-waves "travel much too slowly for electrical waves," Wagner reported. "They seem to be an altogether different entity. That's what makes them so intriguing. They don't seem to be electromagnetic waves at all."

These ideas are far from being scientifically established, and some are certainly fringe. Still, they ought to pique our curiosity about the kinds of questions we have asked about trees, and the kinds we might be asking.

CHAPTER 18
Sitka Spruce

ONCE UPON A TIME, on the banks of the Yakoun River in British Columbia's Queen Charlotte Islands, a sacred three-hundred-year-old Sitka spruce grew in the midst of an old-growth forest. It was a big tree, more than 160 feet tall and six feet in diameter, but it wasn't its size that made it special. The spruce had a genetic mutation that caused its chlorophyll to break down under UV rays, which in turn caused its needles to turn gold. It stood out in a forest of dark green, and some said it seemed to glow from within. In John Vaillant's 2002 *New Yorker* article about the tree, one woman described how, as the sun emerged on a cloudy day, she came upon the tree, and "suddenly there it was in its golden brilliance. We called it the ooh-aah tree, because that's what it made us all say."

The Haida tribe, the aboriginal people who live on the island, believe the sacred tree was a person, one of their ancestors, and they called it K'iid K'iyass, or Old Tree. According to legend, long ago an old man and a young boy were running away from their village, which the Creator had buried in snow as a punishment for the wicked people who lived there. The old man told the boy not to look back at the village, but he did and was turned into the unusual tree.

The Haida are one of myriad cultures across the world and throughout time who have revered a tree as sacred. In ancient Egypt the god Osiris in his earliest expression was believed to inhabit trees, while the Persians believed both good and evil spirits lived inside trees. In Oaxaca, Mexico, a massive cypress, El Arbol de Tule, or the Tule Tree, is venerated, decorated with locks of hair and ribbons, and Native Americans pay homage to many sacred trees with tobacco and other offerings. In Montana's Bitterroot Valley, a massive medicine tree was venerated for generations by Salish, Nez Perce, and other tribes, until someone poisoned and killed it because it stood in the way of a road widening project. "People use places like the Medicine Tree to pray because it reaches up and brings them nearest to the Creator," a tribal elder wrote about it. Siberian shamans believe trees mediate between Earth and Sky and are antennae for cosmic energy, a belief that resonates with Lawrence Edwards's and David Milarch's ideas. Perhaps these galactic energies were the source of the Buddha's enlightenment beneath the sacred fig tree known as the Bodhi tree. "There is little doubt that most if not all races, at some period of their development, have regarded the tree as the home, haunt, or embodiment of a spiritual essence," wrote J. H. Philpot in *The Sacred Tree*. "The god inhabited the tree . . . not in the sense in which a man inhabits a house, but in the sense in which his soul inhabits his body."

Many cultures also believed in a form of the Cosmic Tree, a symbolic tree that grew through all the realms of the world and symbolized the unity between the divine realm, the human realm, and the underground. The Norse people called it Ygg- drasil, an ash tree, which formed the axis of the three levels of the nine worlds of the Norse cosmogony; in Indian cultures the Asvatha is the Cosmic Tree. The Tree of Life is another symbol that appears in numerous cultures and speaks to the bounty of life on the planet, food, shelter, and other gifts for human- kind. It appears in the Garden of Eden, along with the Tree of Knowledge, and in esoteric Judaism it is a symbol for a pathway to God. For the Teutons, Celts, and Druids the oak was a symbol for the Tree of Life, because it often remained alive after it was struck by lightning and was believed to be favored by the gods.

As for the golden spruce, in 1997 a troubled unemployed Canadian forest engineer named Grant Hadwin swam across the frigid Yakoun River to where the sacred Sitka spruce stood and took a chainsaw to it, claiming he was doing so in protest of the industrial-scale logging that has ravaged much of British Columbia. The people in the nearby town of Port Clements were devastated. "It was like a drive-by shooting in a small town," one resident reported. The Haida mourned it as they would an ancestor.

Fortunately the tree was felled in the winter and was dor- mant, which meant it remained viable for a longer time than it would have had it been cut down in other seasons. The cuttings were grafted to the roots of normal spruce seedlings by experts at the British Columbia Forest Service Research Station at Mesachie Lake, and some are still growing at the nursery there. In a private ceremony conducted by Haida spiritual leaders, one of the saplings was planted next to the stump of K'iid K'iyass, or the ooh-aah tree. One can't help but wonder, though: while clones of a tree contain the biological properties of a tree, does its sacred nature live on in its genetic duplicates?

CHAPTER 19
Lost Groves and Wonder Stumps

THE COASTAL REDWOODS are the tallest trees in the world, the tallest of them, Hyperion, topping out at 379.1 feet. As big as they are now, there were, just a century or so ago, much bigger redwoods that were cut down for lumber. Though these supertrees are gone, dozens of their stumps are scattered across parts of Northern California, and in these giant reminders of destruction are the seeds of a unique rebirth.

The redwood plan that David Milarch began when he cloned Gramma, Twin Stem, and Old Blue fell apart after he was told he couldn't use their DNA. Even though the Ancient Forest Archive is a nonprofit organization, federal, state, and other government agencies do not license rights to their properties. That meant that the biggest of all trees—the Stratosphere Giant coastal redwood and the General Sherman sequoia and other

champion trees in federal parks and on other public property—
were off-limits. It seemed for a while that the plan to clone the
biggest redwoods and sequoias had been thwarted.

Then in 2008, Bill Werner found a brand-new website that
belonged to a fellow named Michael Taylor. Taylor is a tree
savant. Tall, thin, and much younger-looking than his forty-five
years, Taylor eats, sleeps, and dreams big trees. He was featured in Richard Preston's book *The Wild Trees* because he and
a friend, Chris Atkins, discovered Hyperion, the tallest of the
redwoods, and he helped redwood researcher Steve Sillett locate
other towering redwoods in California for research purposes.
Jake Milarch followed up with Taylor, sending him pictures of a
giant blue oak and a white oak he had taken on a trip to California. "You're the first hit on my website," Taylor wrote back. "If
you are ever in California, I'll show you where the big trees are."

There are "extreme tree" hunters all over the world, people
who love big trees with a deep and abiding passion, and they
scour the wild—and often not so wild—places on the planet
for them, measuring, recording, and sometimes nominating
them for champion status. Taylor is considered one of the best,
bushwhacking and hiking his way across the mountains of big
tree country in California with his laser range finder, a $3,000
Impulse 200 LR, which is accurate in assessing tree height to
within an inch. He lists all of the detailed data, photos, and other
information on his website, Landmarktrees.net.

Taylor's father, James Searcy Taylor, chairman and CEO of a
private real estate investment company called American Capital
Group, which owns land in California, calls his son enigmatic;
he doesn't care about money, as long as he has enough to live on.

Michael went to school in Arcata, California, and there became a full-time big tree hunter, hiking through the temperate rainforest of California looking for trees. He has found dozens of giant redwoods and, as is his right, he has named them—his finds include Thunderbolt (which had been hit by lightning), Bamboozle, Paradox, and Crossroads. David Milarch calls him the "rain man of trees" because of his brilliance with tree data and his uncanny knack for finding the big ones.

Taylor no longer looks for redwoods—the tallest have all been found he says, with LIDAR (Light Detection and Ranging), an airborne remote sensing technology using lasers that generates incredibly detailed topographic maps. That technology has found two hundred redwoods over 350 feet tall. "The frontier of the redwood is over," Taylor told me. "I've moved on to other tree species." He now searches the forests for trees that can be members of the "eighty-meter club," about 260 feet. He is looking hard for the first known hundred-meter tree (about 328 feet), which seems to be a threshold that few besides the redwoods can crack. He has found a 99.7-meter Douglas fir in Oregon, though the top is dead; and there is a eucalyptus tree in Tasmania that is 99.6 meters that has a live top and will, he says, break the hundred-meter barrier as it grows.

Taylor, who lives in a remote village of some 250 souls called Hyampom, in the redwood country of northern California, finds his trees in two ways. The first is by looking for places where there is water and protection from wind, the two most critical elements for a tree to attain extreme tree status. The second way is by intuition, something he remembers having since he collected snakes as a boy. "The most difficult snake to find was a king snake," he explained. One night, on his way home from an ice cream stop with his parents, he had a clear vision, "like a movie, playing over and over in my head," of a king snake crawling across his driveway. "Ten minutes later, there it was in the driveway." He was so shocked, though, that he didn't react in time to catch it. Now he uses the same sense to find trees. "A lot of the trees I find are through intuition," he said. "It's not psychic, really, but a knowing. I had a powerful dream about Hyperion and then a month later I found it. It's not been found, but I know where the tallest sequoia is, I just know. It's a knowing. The same with a sugar pine. The eighty-meter sugar pine is about to be found." The knowing might be coming from the trees themselves, he allowed. "Trees give off a detectable energy, and they also serve as a grounding for the energy coming down from the stars. I can feel it. It's something that you cannot explain with words."

Archangel had stumbled upon someone who knew more about the big trees of California than anyone else, and who shared their feelings about ancient trees.

ON THEIR NEXT trip to California, Jake Milarch and another grower for Archangel, Tom Broadhagen from Empire, Michigan, met up with Bill Werner and visited Taylor. They in turn introduced him to David Milarch, who began a friendship with Taylor centered on their shared fascination with big trees. The Milarchs told Taylor about the problem they faced because they couldn't collect on government ground. Did he know of any privately owned champion redwoods? Or did he know of any large stumps?

"I know the biggest stump *ever*," Taylor replied, referring to the Fieldbrook Stump. As was the case with cloning champion trees, it seems that no one had ever thought of taking clones from a huge stump. But that, too, made a lot of sense. Taylor was signed on as a consultant.

Jake Milarch, Michael Taylor, Bill Werner, and Tom Broadhagen successfully cloned the Fieldbrook Stump, the remains of the largest coastal redwood that ever lived and one of the five largest stumps in the world. Two weeks later, David returned to California with Jake and Michael Taylor to continue the search for the largest stumps that had ever lived, from trees that had been cut down in the 1890s. For the three men, spending twenty-one days combing Northern California for stumps thirty feet and bigger in diameter was a dream treasure hunt. Taylor stayed up late poring over maps and searching the Internet, looking for places where the crew was likely to find large redwoods or stumps on private land, often just outside redwood parks. When they set out in the morning, he turned on his internal compass. "When I travel with David Milarch, we really find trees," said Taylor. "It's beyond coincidence. I've experienced it again and again. We're driving down the road and we suddenly come upon the crown of a giant tree. It's like there's some kind of internal compass guiding us. It spins around and around, and then locks on. That's how we found them." When they came to a big tree or stump on private land, they would stop. David, in his customary role, would knock on the door of the house on the property, give a hearty hello, ask to take cuttings, have the owner sign a release, and offer to bring the family a cloned tree to plant next to the original. On that trip they found and successfully cloned thirteen giant coast redwood stumps, trees that had been much larger than any coast redwoods alive today.

A year later, on a sunny day in early September, Milarch and Meryl Marsh, the global field operations coordinator for Archangel Ancient Tree Archive, drove a few miles out of Arcata, California, not far from Jedediah Smith Redwood State Park, to show me the Fieldbrook Stump. Marsh is one of the world's few female big tree climbers. A marathon runner raised in Hartland, Michigan, she had in 2009 just quit her job working for a pharmaceutical company and was living in Belgium when she came back to Michigan for a visit and was introduced to Leslie Lee. Lee thought Meryl would be helpful in organizing European projects for Archangel and suggested she go to Copemish to see the grow operation and meet Milarch.

When Milarch met Meryl, he immediately thought her athleticism might figure into a role at the foundation. He showed her a video of climbers on the big redwoods and extolled the big tree project. As Marsh recounts it, the excitement of watching people scale the big trees made her forget about sales—she wanted to climb. Milarch was a step ahead of her. He had already arranged for Rip Tompkins, a former world champion tree climber, to come to Michigan to teach her and Jake to climb trees. "You start working with a climbing instructor next week," he told her, to her delight.

There was no climbing needed for the Fieldbrook Stump, however. It sits in the backyard of a house in the woods, and as we entered the property and headed to the yard, a man's impish face peered out from behind a thick growth of trees and berry bushes. Bill Daley, the owner of the property, was standing on the Fieldbrook Stump, which was overgrown with a tangle of blackberry bushes and small trees. The Fieldbrook Stump, cut down by loggers in 1900, is all that remains of one of the largest coast redwoods that ever lived. It's almost thirty-three feet across at a little above breast height, without the bark, which would have added another two feet to the width. The diameter of the largest living coast redwood is about twenty-five feet with the bark, depending on where it's measured. Crudely hewn steps are still visible on the stump where loggers carved them so they could climb up the tree and pound in springboards, which they stood on as they wielded axes and pushed and pulled the massive two-man saw that was needed to fell the giant.

The Fieldbrook Stump is a local celebrity. There is a black-and-white group photo, taken not long after it was cut down, with eighteen serious-looking schoolchildren in their Sunday best sitting and standing on the stump. Dances were held on it, nuptials exchanged, and children conceived on it. "It was the party place," Daley told us. "Young people brought beer kegs and sat on the stump and carved their initials." These days people who used to live in this region, known as the Redwood Coast, bring their children by to pose for photographs on it, and Daley accommodates them. "My daughter and I counted the rings about twenty years ago for a school science project," said Daley. "The best we could do, because there were sections of the stump missing, is that it was between two thousand and three thousand years old."

There is a two-foot-diameter redwood tree, some thirty feet high, growing out of the base of the stump. It's a sucker, or, technically, a *basal sprout*. Most trees, including redwoods, clone themselves by sending up shoots from the roots of the stump below ground. They become trees in their own right and have the same genetics as the tree that was cut down, but because they are not rooted in the ground and have no root system of their own, they are not as sturdy as the original tree and in heavy storms can be blown away. These sprouts are what Marsh and Milarch were after. And once the clones were viable, they would have their own root systems.

Marsh took a few additional cuttings of the Fieldbrook Stump that day. After she shot a GPS location for the tree and took photos and videos, she pulled out a bag with pruners, sponges, and other gear. She cut about twenty foot- to foot-and-a-half-long needle-covered branches from the base of the enormous sucker and stuffed them in a plastic trash bag. She tossed in some sponges and sprinkled in some water from a bottle, to keep the cuttings moist. The next day she would overnight them to Bill Werner in Monterey, where they would be treated with hormones, placed in rooting medium, and set on mist benches. If all went well and tiny root hairs could be coaxed from the cuttings, in three to four months the cuttings would be cloned and become hundreds of baby redwood trees with the exact same genetics as one of the largest trees that ever lived, one that had been given up for dead. Eventually, through advanced cloning techniques, thousands of copies could be made. "The Fieldbrook Stump is in a class by itself and will live again around the world in the clones we make," David Milarch pronounced. "People have studied it, partied on it, copulated on it, and some probably wept, but we're the first ones who came to help it live on. We can use this tree as a symbol of the reforestation of the world, because we will use its descendants to begin that process. This tree is a symbol that it's time to put things back the way they were." He promised Daley that he would bring him one of the offspring so he could plant it. "We're going to plant super groves of redwoods in England, Ireland, France, Australia, Chile, New Zealand, and of course California," Milarch told me. "We're going to plant four or five thousand seedlings, and for every hundred seedlings we're going to plant one clone of the Fieldbrook Stump and other big trees, so those old-growth genetics can begin to mingle with the other trees. Then we will walk away and leave them alone and let nature do her thing. That will be Archangel's gift to the world. This stump is back from the dead, same as I am. We're both playing the same role. It's a resurrection. This is where we ask the world to help heal itself. We won't read the end of the book, but our grandchildren and great-grandchildren will."

Just a few miles out of Crescent City is Wonder Stump Road, named after the Del Norte County Wonder Stump. We drove along it, keeping our eyes out for a large stump, and we asked the locals where it was, but no one seemed to know. Later, at the Crescent City library, we found an old, faded black-and-white photo of a man in a cowboy hat holding a rifle and sitting on top of a large redwood stump, which is in turn on top of a prone stump beneath it. After some searching we found the Del Norte Wonder Stump on a private piece of ground, covered by small trees and brush, unmarked. The Wonder Stump, it seems, is actually two redwood stumps—a thousand-year-old tree stump whose giant, clawlike roots have grown over a fallen fifteen-hundred-year-old redwood. From its name we had thought it might be larger than most, but it wasn't nearly as large as the thirteen stumps Archangel had already cloned a year earlier. David and Meryl weren't interested in taking cuttings of it.

Throughout the forest here, people have built homes amid the surreal ruins of the world's most dramatic big tree logging. Many homes have giant stumps shaped like the Devil's Tower monument in their front or back yard. Some stumps are covered with vegetation. Others have been incorporated into landscaping, made into playhouses, or decorated with lawn ornaments. After Meryl and David collected more redwood DNA from various stumps, they headed south to rendezvous with Michael Taylor, who, following his intuition, had found a little-known grove of old-growth sequoias on private property. The next day David called to tell me he'd hit the mother lode. They were going to take some samples and return in a month; he asked me to join the follow-up expedition to see some of the biggest trees in the world.

In October 2010 I flew into Fresno and met up with Milarch. The other members of the team were already at the sequoia grove, and he and I set out to join them in a rented SUV. Getting to the Lost Grove, as Milarch dubbed it, was an adventure. We drove through farm country and then climbed a serpentine road into the boulder-studded Sierra Nevada, a wild part of California. In the hills, armed Mexican drug cartels grow marijuana and rattlesnakes live under rocks. At lunch our waitress told us that a mountain lion had been prowling around her house and that she had just secured permission to shoot it. Later we saw a black bear on its hind legs taking apples off a tree in a backyard in Sequoia Crest, one of the most unusual rural subdivisions in the world.

Sequoia Crest, south of Sequoia National Park, was once a 670-acre logging operation owned by the Rouch family. Claud Rouch bought this land in the early 1940s, set up a sawmill, and started logging big trees. The business grew, and at one point there were twenty loggers and seventy-five workers at the mill cutting out the virgin pine, fir, and cedar to build homes in Los Angeles. Incredibly, in spite of their value, the Rouches never touched a sequoia. When I asked Claud's son Sonny, now a vigorous ninety-three years old, why they never cut down the valuable trees, he just smiled and said he didn't know. The property is now a subdivision with a hundred homes, most of them second homes, scattered across a mountainside pell-mell beneath a canopy of giant trees. The Rouch family still owns over five hundred acres of the land, though, including a man-made lake that Sonny built himself and what David Milarch referred to as the Lost Grove of giant sequoias, which is right where it shouldn't be, not below the tree line but at the very top of the mountain. Sonny told us that the grove was considered sacred by Native Americans, and it is here that Sonny wants his ashes scattered.

This Lost Grove is one of the sixty-eight or so groves in the mountains of central California, where the altitude, moisture, and soil come together perfectly to create these big trees. It is the only place in the world they are native.* And Sonny Rouch's trees aren't encumbered by federal or other regulations that

*The number of sequoia groves varies according to who is counting and depends on what defines a discrete grove. Sequoias cover about 35,000 acres, and there is a total of perhaps ten thousand trees.

would prohibit the collection of DNA. When Milarch proposed the cloning, Rouch was eager to have it done.

This grove is critical to the Ancient Tree Archive mission. "Sequoias like moist feet," says Milarch, "but these trees are high and dry. They have adapted to dry conditions without much moisture and at the southern edge of their range." In other words, they could be a critical genotype for life on a hotter and drier planet.

Milarch and his crew drove a four-wheel ATV up the mountain to the grove. There they would take cuttings from as high as they could climb on the sun-exposed side of the tree. To reach those branches, they would access the tree on the lowest branches, at around a hundred feet in the air, and then move up into the crown to take the cuttings. Jake Milarch had rigged up a bow and arrow with a fishing reel and line attached, a contraption he made after he saw big-tree climbers using something like it on the Internet. If an arrow with fishing line tied to it was shot over a branch but missed, it could be reeled in, and he could make another attempt. If the shot was successful, the fishing line would be used to haul up a heavier climbing rope. Here the climbing was different than it was at Roy's Redwoods. The climbers, wearing harnesses and helmets, used ascenders, tools for both the feet and hands, which allowed them to pull themselves up more quickly and with much less effort than the method used with the redwoods.

It takes a certain constitution to stand on a branch at a height that would induce vertigo in most people and reach out and take cuttings, even with a harness. The branches, though over a foot thick, are brittle and can break at any time. Still, Marsh said the height doesn't bother her. "At first I was a little nervous. But I started focusing on what I had to do, and as soon as I did, it wasn't an issue. And I love it. You get an exclusive view that almost no one else gets, just you and the birds. You get to leave everybody on the ground and leave behind whatever drama is going on down there. It's serene and peaceful, and as far as I am concerned, it's the best place to be."

Later the crew drove down a rutted two-track logging road on another part of the property to the Waterfall Tree. This is the largest-diameter single-stem tree in the world—57 feet across and 155 feet in circumference—and the fifth-largest sequoia tree by volume. It grows on top of a steep ravine, near a small waterfall. The thick base of the tree is obscured now because dirt was pushed up around the base when a road was built.

Over the course of a few days, Jake and Meryl took numerous cuttings from the trees, which were then driven down the mountain to be overnighted to Archangel. Like the redwoods, these ancient trees would be a challenge to clone. Old trees don't have the vitality of younger trees. Bill Werner has an 80 percent success rate when he clones a young redwood tree; with the old redwoods he had a success rate of between 4 and 7 percent. Bill Libby was unable to clone sequoias more than eighty years old.

Once Werner received the plant material he followed the usual procedure of disinfecting, selecting, and trimming cuttings to size and wounding the base of each cutting by removing a thin slice of bark to expose the cambium. The cuttings were dipped in a rooting hormone solution and inserted in propagation flats containing a mixture of peat moss and perlite, a growing medium made from volcanic material that increases aeration and drainage. The cuttings must be placed at just the right depth in the soil. "Too deep and you're toast because you get stem rot," Werner explained. The cutting flats are placed on benches with intermittent mist in a temperature- and light-controlled greenhouse. Every twenty minutes, a timed sprayer comes on with a hiss and mists the plants. Werner continually adjusts the hormone ratio in a range from one thousand to eight thousand parts per million, to see which concentration will work. "It comes down to a lot of trial and error, a lot of record keeping," he told me. Once the plant is treated with hormones and set in flats, said Werner, "You wait and pray and water."

Bill Werner and David Milarch have very different spiritual approaches to horticulture. Milarch says he works with a nourishing light—he asks the plant *devas* for pure white light to surround the top of the plant, and he asks for emerald green light from the earth to be held around the roots. "That's where some real magic can happen," says Milarch. Werner's men's Bible study group gathers and prays for, among other things, the plants to grow new roots. The ancient Stagg and Waterfall trees were the subjects of frequent prayer. "I have a certain amount of skill, but I ask for God's favor, whether for healing or a scientific breakthrough," Werner told me. "That includes cloning the Stagg and Waterfall trees." As for Milarch's approach, Werner says, "He acts upon his conviction, that's what's important." Both Milarch and Werner believe that the divine energies they invoke are behind their unusual success in cloning the world's oldest trees.

After three months of mist and prayer, the clones of the ancient sequoias finally sprouted. Milarch called, sounding like a proud father. "We've got roots on the cuttings from the three-thousand-year-old Waterfall Tree!" Then he emailed a photo of the cuttings with the long white beansprout-like roots dangling off the end. Emails of other sequoia and redwood clones followed. No one was more amazed at the cloning success than Bill Libby. "I had to eat crow on the giant sequoias," he said, "because I was convinced it would be very hard to root cuttings over a hundred years old. But Bill Werner has done it to enough of them with a high enough frequency that I had to go swallow that bird."

Clones of the trees are now growing and will soon be copied. In the next year, Milarch says, it will be time to farm out the babies to other countries and begin the creation of supergroves.

CHAPTER 20
Sequoia

FOR CENTURIES CALIFORNIA has had an aura of the fantastic. The state's name comes from the sixteenth-century novel *The Exploits of Esplandián* by Garci Rodríguez de Montalvo, which describes a mythical island adjacent to the gates of a terrestrial paradise. California was rumored to be the site of El Dorado, the fabled city of gold where tribes of Amazon women were said to wander the land. There were tall tales about waterfalls that flowed uphill and a buried core of solid gold, the mother lode. One fantastic story that was true, however, was the tale of the giant trees that grew in the wilderness.

The first known white man to see them was a miner, Augustus Dowd, in 1852. When he returned to camp with his tale of big trees, no one believed him. Later he named the first tree he spied the Discovery Tree, and people started to make pilgrimages to see it. The grandiosity of the trees pierced the hearts of even grizzled frontiersmen. In the 1850s an army surgeon named Lafayette Bunnell rode into a grove of sequoias and later wrote down his thoughts: "It seemed to me I had entered God's holiest temple, where that assembled all that was most divine in material creation."

Some were moved for mercenary reasons. In 1853, when Captain W. H. Hanford, John Kimball, and Ephraim Cutting, all principals in the Union Water Company, finished building a canal, they swung their attention to a grove of sequoias, especially a breathtaking grove with ninety-two big trees, which they saw as a ticket to a career in show business. Their ambitious plan was to drop one of the giants, strip off its bark, wrestle it onto a wagon, take it to San Francisco, put it aboard a schooner, and sail it around the Cape of Good Hope and up to New York City, where they would charge a fee for visitors to see a part of a now dead wonder of nature. The men set their sights on Dowd's Discovery Tree, twenty-four feet in diameter and more than twelve hundred years old.

The Discovery Tree was not, by a long shot, the biggest of the sequoias. Nearby lay a prostrate Father of the Forest, for example, which had died a natural death and fallen sometime before the white men arrived; it was 400 feet tall and 110 feet around, and the first branch was two hundred feet above the base. A fire had hollowed it out, and an early travel writer, James Mason Hutchings, wrote that a man sitting erect on horseback could ride from the bottom into the tree for ninety feet. "At the end of the burnt cavity within, is a never-failing spring of deliciously cool water," he wrote. After a hard rain, a pond formed on the trunk big enough to hold a steamboat.

The gung-ho crew of former canal builders set about their attack on one of the most massive creatures on the planet. They stripped fifty feet of foot-and-a-half-thick bark off the Discovery Tree in ten-foot sections. To fell the tree they drilled three-inch-diameter holes all the way around it into the base with a pump auger, a device they had used to drill out the center of much smaller trees for wooden flumes. They then used saws to connect the drill holes. It took more than three weeks to complete the cut through the trunk. However, when the sawyers finally finished, the tree refused to fall. The men pounded large steel wedges into the cuts in the tree, but it still didn't budge. Two days later, however, a storm rose up and the tree began to "groan and sway in the storm like an expiring giant," according to one witness, until it fell, burying itself twelve feet into the ground and creating a sound like thunder that could be heard fifteen miles away.

The tree was shown briefly in San Francisco, where a local newspaper reported that "thirty two couples waltzed within its enclosure." Then it was placed on a ship bound for New York. Unfortunately for the entrepreneurs, P. T. Barnum turned down the tree, and their ambitions were dashed.

The killing of the Discovery Tree sparked outrage over such crimes against nature. Newspapers carried long, florid descriptions about the felling of the giants. The image of revelers prancing on the remains of the once grand sequoia so angered the conservationist John Muir that he wrote an article titled "The Vandals Then Danced Upon the Stump!"

The outrage helped propel the creation of Yosemite, first as a state park, even though the Calaveras Groves, where the tree had been cut, were, ironically, left out of the park. (In 1931, though, the Calaveras North Grove became Calaveras Big Trees State Park.) But the big tree brouhaha didn't deter another entrepreneur from trying to make his fortune from another giant in the same grove. In 1854, George D. Trask brought in a crew to cut down the Mother of the Forest, a tree 363 feet tall, 31 feet in diameter, and more than 2,200 years old. It was perfectly round and straight as far up as anyone could see.

Trask's approach was to take only the bark, although that would of course kill the tree. His crew set up elaborate scaffolding and peeled sixty tons of bark, in two-foot-wide by eight-foot tall sections, up to one hundred sixteen feet. They numbered the sections of two-foot-thick bark precisely, so the tree could be pieced back together when it went on display.

The *New York Tribune* publisher, Horace Greeley, bought the massive bark and dubbed the tree that remained the "Tree Mastodon." He had it reassembled in New York and London, and it played to sold-out crowds. Mother's naked carcass still stands in the Calaveras Big Trees State Park, with the scaffolding marks visible. News stories about the trees resulted in large numbers of people journeying to see the forest of giants for themselves. Another big tree would not be cut; ecotourism would save the rest.

The giant sequoia, *Sequoiadendron giganteum,* like its relatives the coastal redwood and the dawn redwood, is an ancient species. Its family of trees, Taxodiaceae, dates back 200 to 230 million years to a time of the other giants, the dinosaurs. The range of the sequoias was greatly diminished about a million years ago by climate, leaving it only on the west side of the Sierra Nevada. The groves are widely scattered, which may be the result of the fact that they grew between tongues of a glacier during the Wisconsin period, which began 110,000 years ago and ended 10,000 years ago.

There is concern about the effect of climate change on these goliaths. I asked Nate Stephenson, an ecologist with the U.S. Geological Survey who studies the sequoias, what should be done to protect them. "That's the grand debate, and it is profoundly complex, socially and philosophically," Stephenson said. "Some people say hands off. Some people say assist migration, and some say water them. My hunch is that it will come down to all three. They are all on the table."

CHAPTER 21
A Bioplan

When is the best time to plant a tree? Twenty years ago. The second best time? Today.

CHINESE PROVERB

UNDER A GUNMETAL sky, David Milarch walked around the site where the train station once stood in Copemish. Copemish is an Ojibwe word for "place with the big beech tree," but the big beech, where Milarch's great-grandfather once traded with the tribe, is gone. The village of around two hundred inhabitants is now a melancholy place, and many of the homes are dilapidated. Six generations of Milarchs have lived here.

David showed me the trees he has been planting here, near the converted warehouse that houses the grow operation for the trees—stout six-inch-caliper trees with ninety-inch root balls, maples, white pine, and other natives. The planting is heavy on top of the old train depot, to help remediate the oil-, coal-, and diesel-contaminated soil. "We will rebuild the filter system, clean the air and water and the pollution, and make it habitable for people and for wildlife with these trees," he said. The forest of new trees will connect two fragments of existing hardwood forest, which will provide better cover for deer and coyotes and other wildlife to create a corridor to forests on either side of the road. "It's hotter than Dutch love in the summertime here, and we can cool it with these trees," Milarch said.

A few years ago the town sold dozens of the old maple trees in the city park for $4,000 to someone who wanted to buy them for lumber, and Milarch is replacing those. "They sold trees from the park!" he exclaimed. "That was the straw that broke the camel's back, and I am the camel. I have been all over doing projects to reforest the planet, and I said, 'the next one starts at home.'" He is changing minds, he hopes, as much as the landscape. "Copemish has always been the wrong side of the tracks, but I think people are thinking differently about this place." After seven semi loads of trash were taken from the once decrepit potato warehouse and it was remodeled, it is now bustling with people who are cloning, growing, and shipping trees, and at the other end of the building there is a company that makes solar and wind generators. There are ten or twelve well-paying new jobs in Copemish, people coming and going, and an all-too-rare feeling of prosperity.

Milarch gave me a tour of the grow facility, proud that his longtime dream has come to pass. Giant brilliant lights, bright enough for night baseball, hang from the ceiling illuminating thousands of plants a dozen different shades of green. "At night this place is lit up like a UFO," he laughed. He introduced me to the office manager of Archangel, Marybeth Eckhout, a longtime Michigander who loves her job at Archangel. Marybeth told me that in order to get things done she has to keep Milarch away from the office equipment because of the apparent changes in his electrical nature caused by his NDE. "Good God, he locks everything up," she said. "I've had every electrical piece of equipment—computer, scanners, fax machine—seize up when he's around. Not always—it's worse when he's upset. I have to tell him to calm down."

Collections of big tree genetics have now taken place across the United States and in Ireland, including cuttings there from the Charleville Forest, a private estate that has the massive Charleville King oak on its property, a tree featured in the book *Meetings with Remarkable Trees*. Seventeen of the fifty largest oaks in Ireland grow in Charleville, and Marsh has taken samples from most of them. In the coming year, Bill Libby, David Milarch, Meryl Marsh, and others plan to take genetic samples from the cedars of Lebanon, the massive kauri trees in New Zealand, and a range of other big old trees around the world to create what Milarch calls "the global collection," clones of a hundred of the planet's biggest and oldest trees. While the trees Archangel Ancient Tree Archive has cloned have been done largely by propagating cuttings—setting cuttings in growing medium to root—that method is time-consuming, and the multiplication rates are slow compared to the copy machine of tissue culture, or *micropropagation*. Tissue culture will be key to Archangel's reforestation plans as production ramps up. Once a dozen or two dozen copies of each champion are grown from cuttings, additional copies will be mass-produced by tissue culture.

I toured a fruit tree propagation facility in Canada to see how tissue culture is done. In a brightly lit lab, two workers took a shoot tip containing a tiny meristem from a single tree and dropped it into a jar containing an inch or so of clear gel with growth hormone. A *meristem* is a group of undifferentiated cells at the growing point of a shoot that is capable of develoing into various organs or tissues, such as a new branch. The meristem often has a great deal of vigor. After a few days in the hormones it sprouts into five new green shoots, each of which resembles an alfalfa sprout. A technician pulls the emerald green shoots apart with tweezers and puts each one into another hormone-filled jar. In a few weeks each shoot makes several more, and they are pulled apart and placed in their growth-medium-filled jars, and the number of trees—technically called *plantlets*—grows exponentially. When the plantlets are about an inch long they are transplanted to soil in tiny shells and taken to a heat-and-humidity-controlled greenhouse. Within a year, they will be about six inches tall and ready to be planted in a special soil mixture. In a year's time, tens of thousands of genetically identical trees can be grown from a single meristem from a single tree.

Back in the Copemish growhouse there have been pitfalls along the way, in spite of the good fortune. In 2009, as they were growing the first crop of champions, a black mold called *mucor* infected the warehouse, spreading across the wall and ceiling in clouds of tiny black speckles. "Sixteen thousand rooted cuttings of thirty varieties all died in a month's time," said Milarch. "There was nothing we could do to stop it." It was a huge blow, and Milarch said he almost threw in the towel. They rebuilt, however, cleaning the mold with bleach solution, tearing out walls and relining them with special mold-resistant paneling.

More than seventy species of trees have been propagated under the lights here, among them balm of Gilead, dogwood, willow, and birch. In the back of the warehouse and stored on various other properties there are thousands of cloned champion saplings, ten feet tall or more, waiting to be planted. In California, Bill Werner has hundreds of redwoods, sequoias, Monterey cypresses, and other trees growing around his property, some of them now three or four feet tall. The Brian Boru oak and nearly all of the other oaks from Ireland have been successfully cloned and are between two and three feet tall.

The advice of Diana Beresford-Kroeger, Archangel's science adviser, led the group to focus the first production run on the black willow, one of what they call the waterway trees. These trees will be part of an effort to restore native vegetation to the banks of rivers, lakes, and streams, which will help clean up toxic waste and begin to introduce champion genetics and restore a more appropriate diversity. They hope the genes of the cloned champions will augment the variation present in the wild willows. "They are vital because they are the guardians of our fresh water," Beresford-Kroeger says. "Black willow is the first tree to produce pollen in the spring, which bees bring back to the nest to feed their young, and it has antibiotics and important nutrition to strengthen the brood. The salicylic acid and its daughter compounds, which willows release into rivers and streams, help fish fight infection."

The black willows will be produced and planted with a companion species, the red osier dogwood, with cold-hardy clones

from dogwoods near Yellowstone National Park. Dogwood berries are an important source of food for a range of birds, from
ruffed grouse to woodpeckers to a variety of songbirds, and the
plants also offer cover and nesting sites. Dogwood flowers are
a critical source of pollen for honeybees, while squirrels, chip-
munks, and snowshoe hares eat the twigs. And they can also
help clean the water. "If schoolchildren planted millions of dogwoods and willows in our waterways we could start to clean up
some of the pollution out there," David Milarch says.

The biggest environmental problem of all is the increase of
CO_2 and other greenhouse gases that trap heat in the atmosphere, causing climate change. While the world needs to
quickly phase out the largest sources of carbon dioxide, perhaps the second most important thing we can do right now is
plant trees. Just before this book went to press, a new study by
the Carnegie Institution for Science's Global Ecology was published. Water that transpires from forests, scientists found, has
a robust cooling effect on the planet, not only where the trees
and forests are located, but on the entire global system.

If the right tree—one that will last fifty to a hundred years and
reaches thirty inches in diameter—is planted in the right place,
it will store four to five thousand pounds of carbon over its life.
More important, says Dave Nowak, an expert on the ecosystem
services provided by trees who works for the U.S. Forest Service
in Syracuse, New York, if that tree is planted near a building,
its cooling effect can reduce heating by up to 25 percent. That
means that another 16,000 to 20,000 pounds of carbon is kept
from the atmosphere over the life of the tree because fossil fuels
aren't burned. "If I was going to plant one tree in this country it

would be near a building to reduce energy use, to get both carbon sequestration from the tree and reduction in energy use,"
says Nowak. Placement and type of tree are so critical, though,
that Nowak and his colleagues have created a software program
called i-Tree, in which homeowners can type in their address
and find out energy effects and other services provided when
they plant trees on their property.

Perhaps the next biggest environmental problem the world
faces is the shattering of nature into bits and pieces, which
reduces resilience and greatly challenges nature's ability to
sustain itself, especially in the face of warming temperatures
and other stresses. Again, there's a troubling inverse relationship at work as well: the lack of trees and the fragmentation
and breakdown of natural systems worsen where the human
population grows denser, which is where the need for the ecosystem services that intact native forests and plants provide is
greatest. Eighty percent of the U.S. population, and more than
half of the world's, live in or near urban areas. As the population
grows from six billion to nine billion by 2050, development and
sprawl, and the great unraveling of the natural world, will only
increase.

Restoration forestry should be at the top of the environmental agenda in urban and suburban areas. Merely planting trees as we always have done is not enough—we need a
more radical approach to reforestation and afforestation (the
planting of trees where there weren't any). We shouldn't think
of trees as only beautifying a city or suburb, but as a strategically planted ecotechnology, part of a living, versatile, valuable
environmental infrastructure that cools the urban heat island,

cleans and manages water and air, acts as a natural mood eleva-
tor that reduces anxiety and depression, improves property val-
ues, mitigates noise, provides wildlife habitat, recreation, and
medicines, and grows fruits, nuts, and other nutritious foods.
Instead of trees in the cities, we should be thinking about cities
in the trees. Just as we wouldn't build new developments without proper roads, sewer systems, and other infrastructure, we
shouldn't build new suburbs and other developments without
appropriate forest infrastructure.

For forests to continue to carry out these ecosystem services
into the future, we must create sustainable urban forests. Precisely how best to do this requires study and analysis of both forests and managed landscapes. It's being done in some areas—in
Newark, Delaware, for example, scientists at the University of
Delaware's Center for Managed Ecosystems are looking at how
best to connect some of the fragments in and around the city.
In general it means replanting forests along streams and rivers,
and along the ocean in coastal cities. Forests and trees should
be planted in and around places where people live and should
create corridors to connect to large forest preserves on the edges
of town. The construction of strategically located poplar and willow forests—on rooftops, in and around parking lots, near farm
fields—can treat agricultural and urban runoff before it makes
its way into the oceans as well as provide sewage treatment and
biomass for fuel. The list goes on.

In Diana Beresford-Kroeger's bioplan, honey locusts and
other trees would be planted along roads to absorb pollutants,
and black walnuts could be planted in schoolyards, where aerosols will stimulate the immune systems of children and prevent

cancer while the leaves will take in air pollutants and shield children from UV rays. The walnuts, as well as hazelnuts, butternuts, and other nuts, could be served in school lunches, because they contain omega-3s—nutritionally important oils for brain development—while apples, peaches, and cherries—fruits from other trees that would be planted nearby—could provide healthful food for kids as well. Woodlots would provide wood for shop classes and a source of income for school activities. The motto of agroforestry applies here: "The right tree, in the right place, for the right reason."

The unplanned chaos of the urban and suburban environment, and the fact that most of it is private property, makes it difficult to do large-scale restoration, but education, innovation, and incentives could convince people to roll back the Kentucky bluegrass and turn their lawns and schoolyards into something wild. We can use an alternative approach to agriculture called *agroforestry* that marries forestry and agriculture, to the benefit of both. Trees around or over a farm field can reduce soil erosion and dust and keep the fields cool and more moist. And after the growing season, some trees drop their nitrogen-rich leaves onto the field, where they become fertilizer. Millions of acres of barren land in Africa have been reclaimed with nitrogen-fixing trees. Nuts, fruits, and edible oil crops add to the farm's income.

To ensure the strength of the forests, we should consider including in our urban forests trees with old-growth genetics, to beef up the overall genetic makeup, or add in genetics from trees that have grown in warmer climates, where the trees have adapted to hotter and drier temperatures. In some cases a tree on the southern edge of its range may have adapted to temperatures as much as ten degrees warmer than those of the rest of the forest.

A sustainable urban forest certainly should mean an end to the planting of exotics, or least some types of exotics. While a Bradford pear or a Tree of Heaven might produce the most beautiful flowers or best autumn colors, these French poodles, as David Milarch calls them, are built for a few years of show, not for survival over the long haul, and may not do well as the climate changes. Douglas Tallamy, a professor of entomology and wildlife ecology who studies ecosystems in the eastern United States, says these exotics are hastening the collapse of the biodiversity around us. The majority of insects in this country can't eat the trees most Americans plant. The bugs here evolved to live on oak and hickory and hemlock, and as the trees have disappeared, the insects that eat them have also disappeared. And trees are by far the single biggest generators of biodiversity. Tallamy studied 1,345 plant genera, native and non-native. The top twenty generators of moths and butterflies in the study were trees—first was oak, then cherry, willow, and birch. A native oak species supports 534 species of native Lepidoptera in the mid-Atlantic region alone. On the other hand, ginkgos support four species of caterpillars, dawn redwoods support no species, and Bradford pears sustain just five species. "Ninety-six percent of terrestrial birds eat insects," Tallamy says. With so many exotics, there are far too few insects—too little protein to sustain the birds. "One-third of our bird species are endangered," says Tallamy. "It's a wake-up call." Birds are not the only species affected. The overall deficit of native trees and other plants have led to a decline in a range of species that are beneficial. Exotics aren't always wrong, says Bill Libby, but their use needs to be strategic.

We also need a commitment to research the ecotechnology known as trees. It's unforgivable how little we know about them. They will be on the front lines in a changing world, and we need to study their genetics, ecosystem services, survivability, and other attributes. It's estimated that over a fifty-year lifetime, a tree provides $162,000 in ecosystem services, including $62,000 in air pollution control and $31,250 in soil erosion, and that is a very conservative estimate because things like atmospheric cooling and protection from ultraviolet radiation cannot be valued. Even with this limited valuation, though, we can't afford not to plant trees.

David Milarch, with the help of the Archangel Ancient Tree Archive, is well under way on his quixotic campaign to protect the genetics of the old-growth trees and to create supergroves that will perpetuate the genetics around the world. That idea was his inspiration, and to his credit he has executed it through thick and thin. We won't know for a long time whether he and his vision are correct—if indeed, these genetic qualities are superior or important to the long-term survival of forests. But beyond that, Archangel has joined the growing conversation about, and effort toward, the planting of trees. Planting more trees has been advocated for a long time by a host of organizations and individuals, from American Forests to Trees for Life to Wangari Maathai, the woman who founded the Green Belt Movement in Africa, which focuses on planting trees and women's rights, and who won the Nobel Peace Prize for her efforts. What David Milarch and his colleagues have brought to the discussion is a hard look at what we plant on a changing planet, even if we don't have the science yet to say what is most important, along with a more expansive view of the role of trees and forests.

"Because of the way we have treated the forests, trees—or the lack of them—are the headwaters of all our problems," says Milarch. "Protecting what we have isn't enough, not nearly enough. We need to restore the forests, big time, and help the planet heal itself. It's about restoring the filter system. Instead of spending so much tax money for war and armaments it would be wise to use the army and equipment to reforest the planet as soon as possible. As crazy as that sounds, that is exactly what we need to do, that's the big fix solution. Reforesting and healing Earth is the only answer, and it does a lot more than the science can tell us. That's how we deal with the future—by taking action. Human beings and our machines will not save us. Our government will not save us. And science will not save us. If we wait for their solutions, we will run out of time. But if we harness some of the nearly boundless energy of the planet and the universe by planting trees, it starts into motion the healing and cleansing of the oceans and the atmosphere. And that's what trees do. They benefit every living thing, and it's a gift to our children and grandchildren.

"The big giants of the West Coast, the sequoias and the redwoods, which are so beloved around the world and grow fast, and between them will grow just about anywhere, are the best carbon stackers in the world. They are the ones that will lead us out of this mess. Forty percent of these trees is carbon, and they weigh a thousand tons. That means when they are fully grown they will hold four hundred tons of carbon per tree, and in a forest there are thousands of them, spaced fifteen feet apart. Do the math."

Recently, David was excited about a trip to California, where he had given a talk to NASA, as part of their TEDx speakers series featuring innovative thinkers; he had spoken about the importance of cloning one hundred of the world's landmark trees, planting them in archives, and making their genetics part of the reforestation effort. Some of the scientists from NASA who heard the talk were very interested. "I think it's a fantastic idea," said Steve Craft, the deputy director in the strategic relationships office at NASA's Langley Research Center, who was planning to research it further. "If you are going to plant a tree, why not plant trees genetically you know will survive?" And *Outside* magazine named David one of their innovators of the year in their December 2011 issue.

But there was also troubling news just as this book was going to press. Due to severe budget cutbacks at Archangel, Milarch and his sons were laid off; out of eighteen employees, there are now just a handful. The Archangel Ancient Tree Archive is no longer collecting DNA, though the clones of the big, old trees Archangel has previously collected are still growing and are protected. Milarch the elder continues to volunteer at Archangel in hopes that more funding will come through to continue the mission of cloning the collection of the world's hundred largest trees, and that others, including universities, industry, and government agencies, will take up the task of collection and research. Starting a new chapter is not easy for Milarch, now sixty-two, but he has benefited from minor miracles in the past. Who knows what or who is waiting in the wings to write the next chapter in the story of the big trees.

IF ONE MAN and his family living in the middle of nowhere, with no financial resources, no political connections, no formal education, with little but a vision of making the world a healthier, safer place for his grandchildren and their grandchildren, can bring us closer to a more hopeful and viable future, imagine what we all can do. As scientists admit, science has, for the most part, failed to understand and appreciate the myriad roles of trees and forests even as our planet heads into a dramatically uncertain future. Government has created little or no legislation regarding the planting of trees. It appears that the gap will be filled only by individuals who believe in their—in our—ability to make a difference. The man—or the woman, or the child—who planted trees is potentially everyone.

"*The Man Who Planted Trees* is an illustration of the power each of us has," wrote Frédéric Back, the Academy Award–winning filmmaker who made a film of Jean Giono's book. "If hands and minds come together, we can have an important beneficial effect. We have children and we have grandchildren. Every reasonable person should have a reaction of this kind, to care for the future. It is about the power of work that is life-giving."

We can wait around for someone else to solve the problem of

climate change and the range of other environmental problems we face, from toxic waste to air pollution to dead zones in the oceans to the precipitous decline in biodiversity, or we can take matters into our own hands and plant trees. It may not be the best time to plant a tree, but, as the proverb says, there is no better time.

Notes

5 **Most of these diseases are zoonotic:** Felicia Keesing et al., "Impacts of Biodiversity on the Emergence and Transmission of Infectious Diseases," *Nature*, December 2010.

14 **Los Angeles, for example, has a campaign:** Linda McIntyre, "Treeconomics," *Landscape Architecture*, February 2008.

14 **The floods caused by deforestation:** Corey Bradshaw et al., "Global Evidence that Deforestation Amplifies Flood Risk," *Global Change Biology*, August 2007.

17 **the Wye Oak:** "The Quiet Giant, The Wye Oak," Maryland Department of Natural Resources, www.dnr.state.md.us/forests/trees/giant.asp.

22 **The human-induced drought killed the trees:** "Temperature Sensitivity of Drought-Induced Tree Mortality Portends Increased Regional Die-Off Under Global Change Type Drought," *Proceedings of the National Academy of Sciences*, April 28, 2009.

23 **The most recent forecast for the mean:** "Climate Change Science Compendium," United Nations Environment Programme, September 2009.

23 **by 2100, *half* of all plant and animal species:** E. O. Wilson, *The Future of Life* (New York: Alfred A. Knopf, 2005).

24 **This one is different:** Niles Eldredge, "The Sixth Extinction," action bioscience.org.

27 **By the time the outbreak ends, British Columbia:** "Lands and Natural Resource Operations Fact Sheet," Ministry of Forests, Province of British Columbia, April 2011.

28 **a violent storm swept through the region:** American Geophysical Union, "Staggering Tree Loss from 2005 Amazon Storm," *Science Daily,* July 13, 2010.

28 **In 2010 another, even more severe drought:** University of Leeds, "Two Severe Amazon Droughts in Five Years Alarm Scientists," *Science Daily,* February 3, 2011.

30 **So much forest has died in British Columbia:** "B.C. Forests No Longer a Carbon Sink," report on the Canadian Broadcasting Corporation website, December 9, 2010.

31 **"With many trees now gone":** Richard Macey, "Fewer Trees, Less Rain: Study Uncovers Deforestation Equation," *Sydney Morning Herald,* March 4, 2005.

31 **Trees are also home to an unusual microorganism:** Jim Robbins, "From Trees and Grass, Bacteria that Cause Snow and Rain," *New York Times,* May 24, 2010.

32 **In spite of the lack:** Jim Robbins, "What's Killing the Great Forests of the American West?" *Environment 360,* March 15, 2010.

33 **"These recent die-offs":** Craig D. Allen, "Climate-Induced Forest Dieback: An Escalating Phenomenon?" *Unasylva,* vol. 60, 2009.

35 **But the tree keeps a small strip of living tissue:** Ronald M. Lanner, *The Bristlecone Book: A Natural History of the World's Oldest Tree* (Missoula, Mt.: Mountain Press Publishing, 2007).

35 **In 1964, a graduate student in geography:** Michael P. Cohen, *A Garden of Bristlecones: Tales of Change in the Great Basin* (Las Vegas: University of Nevada Press, 1998).

36 **The study of tree rings from the ancient trees:** "One for the Ages: Bristlecone Pines Break 4,650-year Growth Record," *Scientific American,* November 24, 2009.

37 **Compounding the attack of the bark beetles:** Jim Robbins, "Old Trees May Soon Meet Their Match," *New York Times,* September 28, 2010.

46 **On a warm day in 1970:** A. Kukowka, "The Hazard of the Yew (*Taxus baccata*)," *Zeitschrift für Allgemeinmedizin,* March 10, 1970.

47 **In California's Sierra Nevada:** D. Helmig and J. Arey, "Organic Chemicals in the Air at Whitaker Forest/Sierra Nevada Mountains, California," *Science of the Total Environment,* March 1992.

48 **In French Guiana, researchers sampled compounds:** E. A. Courtois et al., "Diversity of the Volatile Organic Compounds Emitted by 55 Spe-

cies of Tropical Trees: A Survey in French Guiana," *Journal of Chemical Ecology*, November 2009.

48 **The chemicals act in the environment:** Stephen Harrod Buhner, *The Lost Language of Plants: The Ecological Importance of Plant Medicines to Life on Earth* (White River Junction, Vt.: Chelsea Green, 2002).

48 **Studies of the boreal forest:** D. V. Sprackler et al., "Boreal Forest Aerosols and the Impacts on Clouds and Climates," *Philosophical Transactions of the Royal Society*, 2008.

49 **These aerosols play a micro role as well:** Rick J. Willis, *The History of Allelopathy* (New York: Springer, 2007).

52 **Most significant, though, is its role in cancer:** M. N. Gould, "Cancer Chemoprevention and Therapy by Monoterpenes," *Environmental Health Perspectives*, June 1997.

53 **brain tumors that don't respond:** C. Fonseca, "New Therapeutic Approach for Brain Tumors: Intranasal Administration of Ras Inhibitor Perillyl Alcohol," *Neurosurgery* 976, August 2010.

54 **Analysis from the samples taken:** Q. Li et al., "Effects of Phytoncide from Trees on Human Natural Killer Cell Function," *International Journal of Immunopathology and Pharmacology*, October–December 2009, 22(4):951–59.

56 **the effects of green space on children:** University of Illinois at Urbana-Champaign, "Science Suggests Access to Nature Is Essential to Human Health," *Science Daily*, February 19, 2009.

57 **Green spaces, the researchers wrote:** J. Maas et al., "Morbidity Is Related to a Green Living Environment," *Journal of Epidemiology and Community Health* 15, October 2009.

59 **The medicinal properties of the willow:** Daniel E. Moerman, *Native American Ethnobotany* (Portland, Ore.: Timber Press, 1998).

60 **a bitter yellowish crystal from willow bark:** Sunny Auyang, "From Experience to Design: The Science Behind Aspirin," philpapers.org, December 22, 2010.

61 **Aspirin remains something of a miracle drug:** Peter M. Rothwell et al., "Effect of Daily Aspirin on Long-Term Risk of Death Due to Cancer: Analysis of Individual Patient Data from Randomised Trials," *The Lancet*, December 7, 2010.

67 **I'd written a piece about George McMullen:** George McMullen, *One White Crow* (San Francisco, Calif.: Hampton Roads Publishing, 1994).

78 **"It seemed to me I was high up in space":** Carl Jung, *Memories, Dreams and Reflections*, rev. ed. (New York: Vintage, 1989).

78 **A 1992 Gallup poll estimates:** Near Death Experience Research Foundation, www.nderf.org.

79 **Dr. Pim van Lommel, the respected Dutch cardiologist:** Pim van Lommel et al., "Near Death Experience in Survivors of Cardiac Arrest: A Prospective Study in the Netherlands," *The Lancet*, December 15, 2001.

84 **Lumber from old-growth redwoods:** Michael Barbour et al., *Coast Redwood: A Natural and Cultural History* (Los Olivos, Calif.: Cachuma Press, 2001).

88 **Another factor in the survivability of forests:** Bruce Dorminey, "Trees Migrating North Due to Warming," *National Geographic News*, February 9, 2009, news.nationalgeographic.com.

95 **The global tree canopy is a similar story:** H. Bruce Rinker, "Conservation from the Tree Tops: The Emerging Science of Canopy Ecology," actionbioscience.org, October 2000.

96 **DNA is now believed to be mutable:** Ethan Watters, "DNA Is Not Destiny," *Discover*, November 2006.

104 **"Finding a living dawn redwood is at least":** William Gittlen, *Discovered Alive: The Story of the Chinese Redwood* (Berkeley, Calif.: Pierside Publishing, 1999).

111 **"fundamental mistake would be to assume":** Gary Moll, "Trees, Environment, and Genes: In the Evolutionary Battle to Survive and Thrive, a Species' Parentage Is Just the Beginning," *American Forests*, Summer 2003.

111 **she believed Milarch had stolen their idea:** Jim Robbins, "A Tree Project Helps the Genes of Champions Live On," *New York Times*, July 10, 2001.

114 **Jared Milarch and his team set out:** Rick Weiss, "Taking Chips off the Oldest Blocks," *Washington Post*, June 16, 2003.

115 **The Czechs nicknamed the seedling Methuselah Jr.:** "Tree of Life," *Prague Post*, September 13, 2006.

128 **Many people cite the fact that trees create oxygen:** David J. Nowak et al., "Oxygen Production by Urban Trees in the United States," *Arboriculture and Urban Forestry* 2007, 33(3): 220–26.

128 **American urban forests alone sequester:** Linda McIntyre, "Treeconomics," *Landscape Architecture*, February 2008.

128 **Trees also mitigate some obstructive lung diseases:** Gina Schellenbaum Lovasi, "Children Living in Areas with More Street Trees Have Lower Asthma Prevalence," *Journal of Epidemiology and Community Health*, May 1, 2008.

129 **the intact native forest also provides a host:** Gretchen Daily and Katherine Ellison, *The New Economy of Nature: The Quest to Make Conservation Profitable* (Washington, D.C.: Island Press, 2002).

130 **One of the most egregious examples:** David A. Farenthold, "Failing the Chesapeake Bay," *Washington Post* series, December 2008.

131 **A study of sixteen tributaries:** Bernard W. Sweeney et al., "Riparian Deforestation, Stream Narrowing, and Loss of Stream Ecosystem Services," *Proceedings of the National Academy of Sciences*, September 28, 2004.

132 **poplar and willow trees to remediate toxic waste:** Andrew Revkin, "New Pollution Tool: Toxic Avenger with Leaves," *New York Times*, March 6, 2001.

134 **the city has replaced the plant with a willow field:** I. Dimitriou et al., "Willows for Energy and Phytoremediation in Sweden," *Unasylva* 221, vol. 56, 2005.

136 **The importance of iron led one path-breaking scientist:** T. Kawaguchi, "The Influence of Forested Watershed on Fisheries Productivity: A New Perspective" *Thalassas* 2003, 19(2): 9–12.

138 **There may be yet another reason to value the tree:** G. Strobel et al., "The Production of Myco-Diesel Hydrocarbons and the Derivatives by the Endophytic Fungus *Gliocladium roseum*," *Society of General Microbiology*, September 2, 2008.

141 **much of northern Europe could be at serious risk:** Robert B. Gagosian, Woods Hole Oceanographic Institution, "Abrupt Climate Change: Should We Be Worried?" Talk given at World Economic Forum, January 27, 2003.

142 **diseases could wipe out stressed forests:** Emily Beament, "Now Britain's Oaks Face Killer Disease," *The Independent*, April 28, 2010.

144 **The forest and the native culture:** William E. Klingshirn, *Caesarius of Arles: The Making of a Christian Community in Late Antique Gaul* (Cambridge, U.K.: Cambridge University Press, 1994).

144 **one of the planks in the Sinn Féin platform:** Paul Gallagher, "What Is Sinn Fein? The American System vs. British Geopolitics in Ireland," *American Almanac*, January 9, 1995.

144 **they started to mine the vast oak woods:** Rebecca Solnit, "The Lost Woods of Killarney—Old-growth Oak Forest in Ireland," *Sierra* 29, August 2011.

146 **Shortly afterward, in what seemed a miracle:** Gordon Deegan, "Clare Fairy Tree Vandalised," *Irish Times*, August 14, 2002; "Fairies Defy Chainsaw Attacker to Sprout New Leaves on Thorn Bush," *Irish Times*, June 2, 2003.

149 **native forests across the Continent:** "Global Forest Resources Assessment," Food and Agriculture Organization of the United Nations, 2010.

155 One of the largest and most beautiful exceptions: "Kingley Vale National Nature Reserve," Wikipedia.

157 the teachings of the Austrian mystic Rudolf Steiner: Lawrence Edwards, *The Vortex of Life: Nature's Patterns in Space and Time* (Floris Books, 2006).

158 Other scholars have noted a link: Otto Janik and Leonid B. Starosielski, eds., "Supernovae und Baumwachstum" (Supernovae and Tree Growth), in *Exakt: Exklusiv-Informationen aus Wirtschaft, Wissenschaft, Forschung und Technik in der Sowjetunion* (*Exakt: Exclusive Information from Industry, Academia, Science, and Technology in the Soviet Union*) (Stuttgart: Deutsche Verlagsanstalt, 1976).

158 "consistent and statistically significant relationship": Matt Walker, BBC, October 19, 2009.

163 In 2009, researchers at the University of Washington: Roberta Kwok, "A Light in the Forest: Wireless Sensors Draw Energy from Trees," *Conservation*, January 2009.

166 In John Vaillant's 2002 *New Yorker* article: John Vaillant, "Letter from British Columbia: The Golden Bough," *The New Yorker*, November 4, 2002. Readers who would like to read more about the golden spruce can read John Vaillant's book *The Golden Spruce: A Tale of Myth, Madness, and Greed* (New York: W.W. Norton & Company, 2005).

180 This Lost Grove is one of the: "List of giant sequoia groves," Wikipedia.

185 the tale of the giant trees that grew: Joseph H. Engbeck, *The Enduring Giants: The Epic Story of Giant Sequoia and the Big Trees of Calaveras* (Berkeley, Calif.: California State Parks, 1976).

195 Water that transpires from forests: George A. Ban-Weiss et al., "Climate Forcing and Response to Idealized Changes in Surface Latent and Sensible Heat," *Environmental Research Letters*, 2011.

200 over a fifty-year lifetime, a tree provides $162,000: Kathleen Alexander, "Benefits of Trees in Urban Areas: Colorado Trees," which cites David J. Nowak, "Benefits of Community Trees: Brooklyn Trees" (USDA Forest Service General Technical Report, in review).

Index

Discovery Tree, 186–87
disease. *See* health
DNA, tree
and epigenetics, 95–96
library of knowledge within, 97
Luc Montagnier's theory, 162
preserving for champion trees, 7, 13–14
dogwood trees, 194–95
Dow Botanical Gardens, Michigan, 43
Dowd, Augustus, 185–86
drought, impact on trees, 22–23, 24, 29
Druids, 144, 168
Dutch elm disease, 43, 139

Eastwood, Clint, 67, 115
Eckhout, Marybeth, 192
Ecolotree, 132, 135
ecotechnology, trees as, 5, 132, 137, 196–97, 200
edge effects, 87–88, 89
Edwards, Lawrence, 156–58
Ehrle, Elwood "Woody," 41, 43
elm trees, 42–43, 114
Emerson, Norman, 67
endophytes, 139–40
epigenetics, 95–96
The Ewok Adventure (movie), 91
exotics, 199–200
"extreme tree" hunters, 171–73
exudates, 94, 132

farmland issues, 130–31, 134, 135
fertilizers, 130, 131, 134, 135, 136
Fieldbrook Stump, California, 174, 175–78
filters, trees as, 126–36, 201
Findhorn, Scotland, 160
Fischer, Rainer, 164
Flanagan, Martin, 7, 8
flooding, role of deforestation in, 14–15
Florida Champion Tree project, 107–10
Fonseca, Clovis O., 53
Ford, Edsel II, 113
Ford, Henry, Jr., 113
forests
and climate change, xii, xiii, xv–xvi, 22–23, 24, 26–30
die-offs, xvi, xvii, 22, 23, 25–33
edge effects, 87–88, 89
fragmenting, 6, 87–88
and insect infestation, xii, xiii, xv–xvi, 26–27, 29
loss of champion trees, 12, 13–14

natural migration, 88–89
role in natural world, xvii, xviii
as watershed filters, 126–36
See also trees
foundation species, 37–38
fragmenting forests, 6, 87–88

Gangloff, Deborah, 111
Garden Club of America, 118
Garden of Eden, 168
General Sherman sequoia, 170
genetics
cloning champion trees to protect, 7, 8, 12, 13–14, 41, 82–83, 93
Milarch begins collecting for champion trees, 41–44
seedlings vs. clones, 7, 93, 114–15
giant sequoia (*Sequoiadendron giganteum*), 101, 188–89
Giono, Jean, 3, 203
Gouin, Frank, 8–9, 17, 18, 19–20, 93
Gould, Michael, 53
Great Basin National Park, Nevada, 35–36
Greeley, Horace, 188
greenhouse gases. *See* carbon stores

Hadwin, Grant, 168
Hageneder, Fred, 153, 155, 160, 161–64
Hanford, Capt. W. H., 186
Hansen, James, 23
Harvey, Paul, 43
hazel trees, 148, 198
health
role of forests in transmitting pathogens, 5–6
study of effects of trees on humans, 52–58, 139
Heisenberg, Werner, 158
Hippocrates Tree, 121
Hu Hsen-hsu, 102
human-assisted tree migration, 104, 125, 142–43
Hutchings, James Mason, 186–87
Hyperion (coastal redwood), 170, 171, 173

imaginal realm, 80, 159
insects
epigenetic responses by trees, 96
impact of climate change on infestations, xii, xiii, xv–xvi, 26–27, 29
impact of exotic tree planting on biodiversity, 199

216   INDEX

Tallamy, Douglas, 89, 199
Taylor, James Searcy, 171
Taylor, Michael, 171–74
temperature
    forecast for mean global rise, 23
    impact of climate change on trees,
        22–23, 24, 26–30
    increasing averages in Rocky Mountain
        West, 29
terpenes, 48–49, 52–53
Teutons, 168
The Spirit of Trees (Hageneder), 162–64
The Wild Trees (Preston), 171
3M Corporation, 115
tissue culture, 192–93
Token, Boris P., 49
Tompkins, Rip, 175
Torreya Guardian, 124
Torreya taxifolia. See stinking cedar trees
toxic waste, phytoremediation, 128, 131–35
transpiration, 24, 31, 195
Trask, George D., 188
tree climbers, 91, 98, 181–82
Tree of Heaven, 199
Tree of Knowledge, 168
Tree of Life, 168
Tree Register of the British Isles, 152–53
Tree Trust. *See* National Tree Trust
trees
    aerosols in, 45–55
    chemical communication, 50–51
    cloning to save genetics, 7, 8, 12, 13–14,
        41, 82–83, 93
    connection to human health, 5–6,
        52–58, 139
    definition of champion, 7
    definition of cloning, 7
    as ecotechnology, 5, 132, 137, 196–97,
        200
    electrical properties, 162–65
    and energy use, xix, 133, 196
    epigenetics, 95–96
    global canopy, 51, 64–65, 95
    moving, 101–4, 125, 142–43
    mystical links, 156–65
    role in mitigating pollution, 128–29,
        131–35, 200
    role of planting, 4–6, 200, 202
    root systems, 48, 94–95, 132, 135
    sacred aspects, xix, 166–69

seedlings compared with clones, 7, 93,
    114–15
self-pollinating, 115, 125
See also forests
Trees for Life, 200
Tule Tree, Mexico, 167

ulmo trees, 138–39
Union Water Company, 186
urban forests, 15, 99–100, 128, 196–99
U.S. Capitol, 113, 114
U.S. Robotics, 119

Vaillant, John, 168
van Lommel, Pim, 79
viral deseases, as ecological problem, 5–6

Wagner, O. E., 165
walnut trees, 48, 119, 197
Washington, George, 20, 113, 115
waste, phytoremediation, 128, 131–35
water management, xviii, 30, 31, 127–37,
    162
Waterfall Tree, 182, 183, 184
Werner, Bill, 99, 114, 171, 173, 174, 182–83,
    194
willow trees
    Archangel propagation, 127, 194
    as filter, 126–36
    as generator of biodiversity, 199
    in Ireland, 149
    medicinal properties, 59–60
    role in remediating waste, 128, 132–35
    usefulness, 59–62
Wilson, E. O., 23, 45, 58
Wolfowitz, Paul, 112
Woodlanders Nursery, Aiken, South
    Carolina, 125
Wye Oak, 9, 17–20

yew trees
    characteristics, 151–53
    crafting longbows from, 154–55
    Dr. Kukowka's experience, 46–47
    healing properties, 154
    in Ireland, 145–46, 150
    as poisonous, 154
    as sacred, 152, 153–54
Yggdrasil (Norse mythology), 168
Yosemite National Park, California, 188

ABOUT THE AUTHOR

JIM ROBBINS has written for *The New York Times* for more than thirty years. He has also written for magazines including *Audubon, Condé Nast Traveler, Smithsonian, Vanity Fair, The Sunday Times,* and *Conservation*. He has covered environmental stories across the United States and in far-flung places, including Mongolia, Mexico, Chile, Peru, the Yanomami Territory of Brazil, Norway, and Sweden.

Robbins is the author of *Last Refuge: The Environmental Showdown in the American West* (1993) and *A Symphony in the Brain: The Evolution of the New Brain Wave Biofeedback* (2000). He is also the co-author of *The Open-Focus Brain: Harnessing the Power of Attention to Heal Mind and Body* (2007), about the critical and overlooked role of attention, and of *Dissolving Pain* (2010), about the role of attention in pain.

He has lived in Helena, Montana, for thirty-five years.

www.jim-robbins.net.